PISGAH ASTRONOMICAL RESEARCH INSTITUTE

AN UNTOLD HISTORY OF SPACEMEN & SPIES

CRAIG GRALLEY

Published by The History Press
Charleston, SC
www.historypress.com

Front cover: Photograph courtesy of Tim Reaves, timothyreavesphotography.com.
Back cover: Aerial photograph of PARI courtesy of Don Cline. The small back cover photograph—the first full-color image of the entire Earth—was taken by the *ATS-3* satellite, commanded from NASA's Rosman station, on November 10, 1967. This image was on the front cover of the fall 1968 edition of the *Whole Earth Catalog. Photograph courtesy of NASA.*
Opposite: Don and Jo Cline at Don's Rock, PARI. *Courtesy of PARI.*

First published 2023

Manufactured in the United States

ISBN 9781467152181

Library of Congress Control Number: 2022949586

PISGAH ASTRONOMICAL RESEARCH INSTITUTE

To Don and Jo, "the first lady of PARI,"
for your vision and selfless commitment to science education.

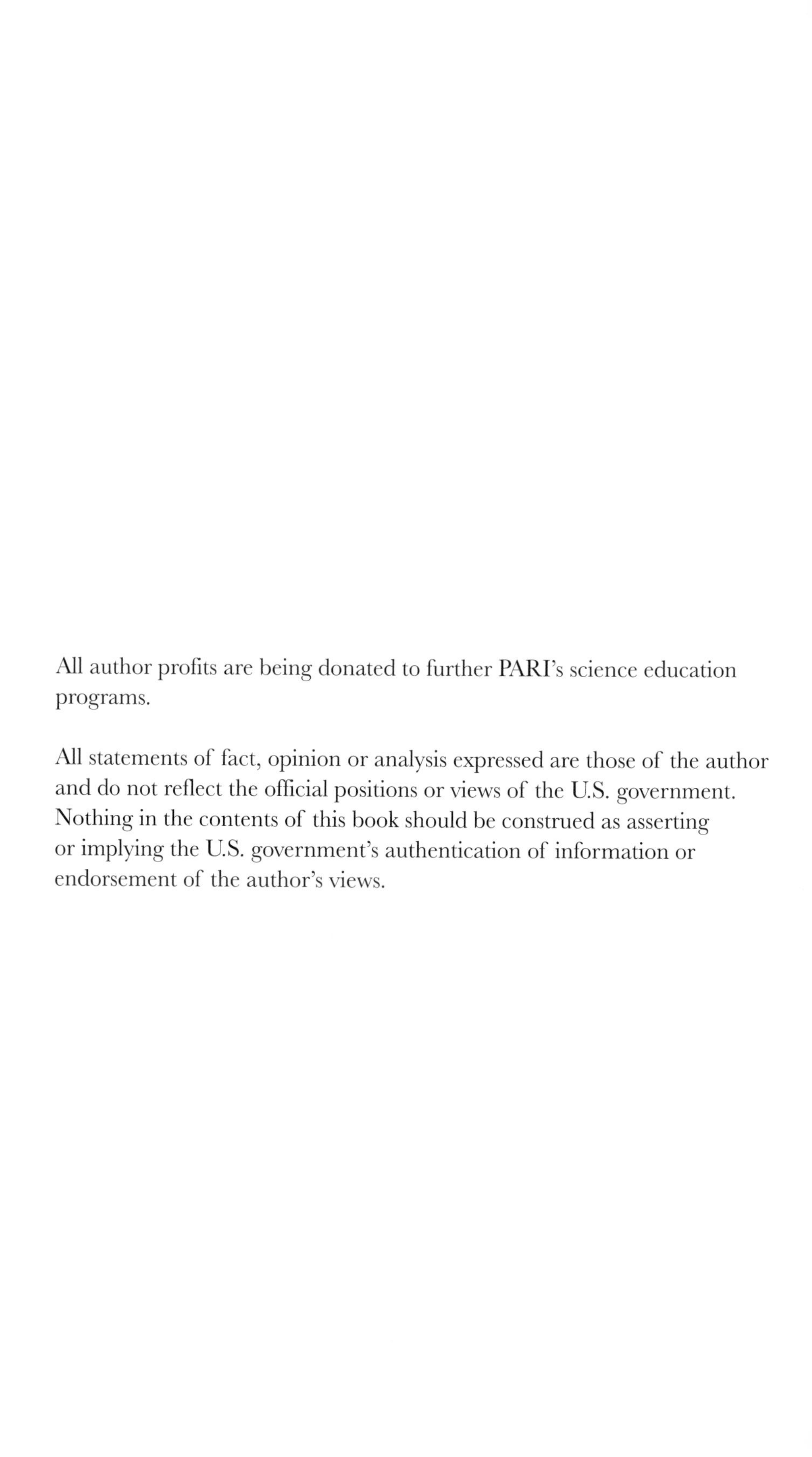

All author profits are being donated to further PARI's science education programs.

CONTENTS

PREFACE

It's a daunting challenge to write a book that includes an organization of the federal government so secretive that its own employees call it "No Such Agency." It was 2008, over a decade after Rosman Research Station (RRS) closed its doors, when the National Security Agency (NSA) lifted the veil of secrecy just a bit and admitted publicly that the Rosman ground station was, in fact, one of its own. Before 2008, the RRS's affiliation with the National Security Agency was classified as "secret." Since then, the information spigot has been completely shut off. No additional NSA documents on Rosman have been released. To be fair, document declassification is a daunting task, and the NSA, like other members of the intelligence community, is overwhelmed by requests to open its vault to the public. So, it prioritizes documents for review and release. Rosman's documents are not at the top of that list. Still, other likely tenants at the NSA's RSS, like the National Reconnaissance Office (NRO), which develops and launches spy satellites, have also been tight-lipped. When asked in a Freedom of Information Act request to verify its presence at the Rosman site, the NRO offered a terse, single-sentence response: "We can neither confirm nor deny the existence or nonexistence of records responsive to your request."

I was concerned. Lacking direct information, how could I hope to write a complete history of the Pisgah Astronomical Research Institute (PARI)? What would I say about the NSA years (1981–95)? Still, my curiosity pushed me forward, and I began researching the National Aeronautics and Space

Administration (NASA), the first "owner" that called the site, awkwardly, the Rosman Satellite Tracking and Data Acquisition (STADAN) Station. Then the pieces started falling into place. I learned about NASA's antennas and its ability to follow satellites in unique orbits. I found the frequencies of four satellite dishes in a sales brochure the Department of Defense (DoD) offered to lure a new tenant to the site. And the Central Intelligence Agency's (CIA), NRO's and NASA's websites offered declassified documents that told of programs NASA pursued jointly with these agencies and programs that Rosman participated in. All of this unclassified information, when pieced together and placed in a historical context, provides a compelling story of how the NSA likely used its capabilities at Rosman and the targets it pursued to successfully protect our country.

This book allows PARI to reclaim its proud history, a history that is much broader than that of any one organization. If a researcher's focus is too narrow, they miss the larger themes of adaptation and achievement that slice through the NASA, NSA and PARI years. Listening to stories and reading about the work conducted at Rosman, I couldn't help but be awed by the dedication of ordinary citizens to accomplish extraordinary things. NASA Rosman engineers commanded satellites that achieved little-known "firsts" in weather forecasting that saved thousands of lives; pioneered global positioning technology that guides us today, rain or shine, safely to work and home; and were at the forefront of developing cellular communications that keep us close to colleagues, family and friends. The accomplishments of the NSA's Rosman Research Station, while less visible, are no less remarkable. No price can be placed on maintaining peace and freedom. The Soviet, Cuban and other communications Rosman personnel intercepted helped our leaders successfully navigate one of the most dangerous and turbulent periods of the Cold War. And today, PARI's work is groundbreaking and revolutionary in its own way, creating a new model for science education that will help our country compete in a technologically advanced world.

This book chronicles the little-known story of a small, rural community, nestled in the woods of western North Carolina, that had an outsized role in our nation's history. It's also a story of great pride. Some employees who called the Rosman NASA or NSA site home over a half century ago still reside in Transylvania County, and most, because of their advancing age or government secrecy agreements, remain silent. In learning more about PARI's history, readers will understand the honor and accomplishment these men and women felt for their work and will hopefully also feel the satisfaction that comes with living in a community with an important past.

This story could not have been told without the help of many others. The first who come to mind are Don and Jo Cline, who had the vision to see what a collection of 1960s satellite dishes could become—an amazing astronomical research institute dedicated to science, technology, engineering and mathematics education for our country's young women and men. They gave freely of their time, talent and treasure for a higher ideal, one that our country continues to underfund: science education. With the recent passing of his wife, Jo, Don carries on with this remarkable experiment as a pioneer in science education, for which there is no precedent or model. I appreciated our many conversations about PARI's past, present and future, and I look forward to a continuing association.

The most interesting books about institutions invariably are rooted in the stories about people. I'd like, in particular, to thank longtime PARI chief technology officer Lamar Owen and chief facilities and security officer Brad McCall for their knowledge, insights and personal reminiscences. Both were early and full supporters of this project. Quite simply, this book could not have been written without their willingness to tell PARI's story. I'm grateful to Jo Ann Jackson, the spouse of the second NASA director at the Rosman Tracking Station, who spoke of her husband, Chuck, and his deep commitment to Rosman's workforce and space research. I'd also thank Eugene "Joe" Collins, senior engineer for the advanced technology satellite (ATS) program at Rosman, who operated the most cutting-edge communications satellite of its day, and former Pisgah Forest district manager Art Rowe, who with longtime NASA, DoD and PARI employee Thad McCall, kept the facility from being shut down after the DoD's departure. Bill R. helped me understand the importance of continuity, as the site changed hands from NASA to the DoD. Photographers Tim Reaves, Doug Tate and Alex Armstrong made outstanding photographic contributions. Alex, one of PARI's army of volunteers, also provided expert assistance in preparing site photographs for the book. Thanks also go to current PARI employees Charles Willard, Ann Daves, Sarah Chappell, Melanie Crowson, Tim DeLisle, Chelena Blythe and James "Buster" McCall. Volunteers form the backbone of PARI, from greeting visitors to maintaining the grounds and satellite dishes and serving as camp counselors. Thurburn Barker, who oversees the impressive photographic archive of astronomical glass plates; Bob Hayward, who, for years, has run PARI's planetarium; and twenty-year veteran Joe Phillips—all volunteers—deserve special mention. Thank you for your support. PARI board members John Avant, Ken Jacobson and Zac Engle provided keen insight into the opportunities that lay ahead for PARI, and former employees Mike Castelaz and Christi Whitworth told of PARI's formative years.

Randi Neff of the Smoky Mountain STEM Collaborative and Mayra Lebron of the University of Puerto Rico highlighted the importance of institutional associations. Matt Shelby, Ryan Barlow and Nathan Whitsett offered their experiences as PARI interns. And Patricia Craig effectively demonstrated how student education at PARI can lead to a rewarding and successful career in science.

Thanks go, too, to the scholars who responded to my requests for information. In particular, the University of North Carolina–Asheville's (UNC-A) Jackie Langille provided expert guidance on PARI's geology and explained why the site is ideal for transmitting and receiving signals. Trying to untangle the land's chain of ownership before the U.S. Forest Service and then NASA came to occupy it was especially challenging. I was fortunate to have the assistance of David Whitmire. Biltmore historian Scott Shumate was particularly helpful, as was my good friend and neighbor, Brevard lawyer Don Jordan. Karin Smith and Delia McCall in the Office of the Register of Deeds for Transylvania County lent a hand, too.

Laura Gardner, Joyce Ralston and Joe Russo of the Transylvania County Library's North Carolina Room were knowledgeable and generous in opening their PARI, Vanderbilt and Silversteen paper and photograph archives.

I'm also grateful to other members of the Transylvania County community who lent a hand in facilitating the completion of this book, including Steve Womble, former Brevard College president David Joyce and the friends and colleagues of the Transylvania Writers Alliance, especially Elaine Hills, Betty Reed, Peter Margolin and Kristin Leesment, who offered comments on portions of the book. The History Press authors Susan Lefler and Marci Spencer provided insights on their research and thoughts on working with the publisher. Thanks, too, to The History Press acquisitions editor Kate Jenkins for her expert guidance.

I was fortunate to have the assistance of Robert Simpson and Spencer Allenbaugh of NSA's National Cryptologic Museum, who helped secure available information through the Freedom of Information Act and a photograph of the original Rosman Research Station sign. I'd also like to thank the Association of Former Intelligence Officers for their support in identifying former RRS employees for this book. NASA's Holly McIntyre and NASA/Goddard's Sarah LeClaire directed me toward information about the organization's early scientific satellites. The National Archives' William

Wade and Kevin Quinn provided expert advice and several excellent early photographs of the Rosman Satellite Tracking and Data Acquisition site while it was under construction.

Employees of the U.S. Intelligence Community are inherently wary of anyone who asks questions about their work, and former Rosman Research Station employees who still reside in western North Carolina are no different. Even though we belong to the same "intelligence fraternity," they politely but firmly deflected conversations that even lightly touched on their employment, and in doing so, they upheld the trust that was placed in them by their former employer. Those who did open the door a crack spoke obliquely about life at Rosman during its DoD years, omitting the site's missions or capabilities. I'm indebted to former DoD employees Art T., Joe K., Bill R., Fred G., and Bruce G. for their insights into life at Rosman Research Station. I am also grateful to friends, former colleagues and science experts whose opinions I value and who generously gave their time to offer thoughts on how to improve the draft. This list includes Michael Roosevelt, David Kelly, Patricia Craig, Greg Colgrove and Lamar Owen. In a speculative piece like this one, there are bound to be omissions and errors in judgment and fact. All shortcomings are entirely my own.

Finally, writing a book is a fulfilling but lonely adventure. I'm fortunate to have an understanding wife, Janet, and a son, Will, who accept my passions and limitations. I appreciate and rely on their sustaining love.

Craig Gralley
Pisgah Forest, NC
www.craiggralley.com

INTRODUCTION

January 1981: the last moving van rumbled down the lazy country byway called Macedonia Church Road. The chain link fence was now locked. Anyone who noticed the brown and white sign by the entrance would say this place was too far from town, buried too deeply in the woods, too distant from anything to be important. But they would be wrong.

The ground, carved from the Pisgah National Forest seven miles outside of Rosman, North Carolina, was once a hive of activity for 260 NASA administrators, consultant engineers and scientists. But now, all was quiet. The scientific satellites the site once proudly commanded that pioneered space science, weather prediction, global communications and uncovered the Earth's natural resources were silent—they either burned up in the Earth's atmosphere or were disabled and left to circle the heavens. The two huge dish antennas, once the cutting edge of technological innovation, were pointed straight up, no longer searching for signals from space.

When men landed on the moon, America had crossed the finish line set by President Kennedy. But NASA's budget eventually dwindled, and spaceflight became an afterthought. What's more, key missions of ground stations and tracking satellites collecting data could now be accomplished in space. After nearly two decades of faithful service, NASA abandoned the Rosman Space Tracking Station.

But its story wasn't over.

Just as NASA was departing, a group of men in dark suits got out of their unmarked van and stared straight up at the two big dishes. Then more people arrived. The perimeter fence was upgraded, cameras and motion sensors were installed and a small army of uniformed men with automatic weapons began to patrol the grounds, day and night. Public tours, which were once allowed, were forbidden, and the guard station was fortified with barricades and barbed wire. Satellite dishes were dusted off, and new ones were erected. A pistol range was established, and buildings were expanded, some outfitted with bullet-proof glass. The secretive new tenants gave the site a bland new name: the Rosman Research Station.

But one building, just constructed, offered a clue to the site's new mission. They called it Building 14, the Shredder Building. It housed a monster with razor-sharp teeth and an insatiable appetite for paper. Tucked away in a corner of the site, it was the final stop for all secret documents, records, notes and scraps of paper. The mechanical beast was fed by a man who couldn't read. All day long—and sometimes into the night—he shoveled paper, mountains of paper, into the creature's gaping mouth. On the other end emerged thin ribbons that couldn't be pieced together. That was the idea.

For the next fourteen years, this site's existence was denied, and its purpose was shrouded in secrecy, hidden behind government reports with strange codewords stamped in thick black letters: TOP SECRET UMBRA. Its mission was the National Security Agency's mission: to collect secret conversations and military signals from the satellites of those who would do the nation harm. The diamonds it found amid the clutter of useless noise were assembled and sent to the president and other top leaders. But then, suddenly, like those who came before, the tenants vanished.

Still, the decades-long story of the Rosman site, which had been through much and accomplished great things, was not over.

Miraculously, in 1998, new tenants arrived and brought the facility back to life. Once again, the satellite dishes were dusted off, but this time, they were pointed toward the deepest stars in space. The newly renamed Pisgah Astronomical Research Institute (PARI), like its predecessors, opened with a brand-new mission: to uncover the secrets of the universe and prepare the country's young scientists to succeed in a new highly technological world.

Like all great institutions, PARI's story is America's story. With each iteration, PARI evolved to fit the needs of America's greatest challenges: space exploration and winning the space race, protecting our nation's security and, today, joining America's quest for scientific research and educational advancement.

Still, despite its rich history, PARI's past is largely unknown. Some claim government secrecy is to blame, but the site has also been, paradoxically, a victim of its own transformation. Different owners, names and missions make for a complex story. The common thread is science and how it helps uncover the many secrets found in the sky and space above. PARI's evolution is unlike that of any former government facility. But through research and declassified government documents, decades-long mysteries are being uncovered, and PARI's history of adaptation and accomplishment can finally be told.

Now in its sixth decade, when most of us think of kicking back for a relaxing retirement, PARI is still working and evolving, focused on the future issues that matter most to this country.

ACT I.

SPACEMEN

THE NATIONAL AERONAUTICS AND SPACE ADMINISTRATION ARRIVES IN ROSMAN (1963–81)

"THE RACE FOR CONTROL OF THE UNIVERSE HAS STARTED"[1]

The year was 1957, and the United States and the Soviet Union, once allies in World War II, were now fierce competitors. It was capitalism versus communism—one rooted in individual freedom and the other in state control.[2] The Cold War could turn hot at any moment. The countries of eastern Europe had fallen like dominoes under the domination of the Soviet Union. Bulgaria, Poland and Hungary were first. The U.S. government called them "captive nations." East Germany, Czechoslovakia and Yugoslavia were next. It was an intense struggle between the world's two dominant powers, each vying to bring other countries into its camp.

The United States–Soviet competition for allies and prestige took a turn upward, toward the heavens, when a completely different frontier opened: space. On October 4, 1957, the Soviet Union launched a 180-pound, beach ball–sized metal sphere into a low-Earth orbit, and it circled the world for twenty-one days, transmitting a rhythmic *beep* to Soviet military ground controllers several hundred miles below.

Though primitive by today's standards, *Sputnik* ("fellow traveler") captured the imagination of people around the world and proved that a communist government could produce stunning technical achievements.

Then one month later, the Soviets launched *Sputnik-II*, this time carrying a stray dog from Moscow's streets named Laika as its passenger.[3] Closer to home, the two Sputnik launches shook America's confidence. The score was: the Soviet Union, 2; the United States, 0. How could the Soviets launch two satellites without a similar U.S. success?[4]

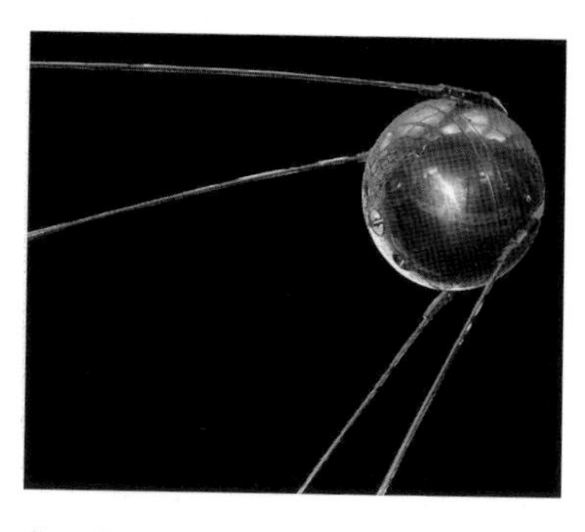

Sputnik's rhythmic beep was heard by amateur radio operators but was not tracked by the U.S. military until a week after the satellite's launch. *Courtesy of NASA.*

With the launch of the Sputnik satellites, a series of troubling questions began to arise. The launches exposed a gaping hole in America's defense, specifically the nation's inability to find and track foreign satellites. Amateur radio operators could hear *Sputnik-I*, but the U.S. military couldn't track it initially because the satellite's transmitting frequency was out of the range of its brand-new minimum tracking ("Minitrack") network of twelve ground stations spread across the east coast of the United States and the west coast of South America.[5] The inability to track *Sputnik-I* had serious implications. If a Soviet rocket could blast off carrying a satellite, some reasoned the next one might carry a nuclear warhead.[6] To understand the nuclear threat from Soviet ballistic missiles and to counter a possible sneak attack, President Eisenhower tasked the CIA with developing the high-altitude U-2 spy plane and, later, photographic reconnaissance satellites, which would be less vulnerable to being shot down.

Meeting the challenge of the space race and understanding the Soviet military threat would consume the attention of U.S. leaders for decades to come. Both tasks would eventually fall on the shoulders of those who worked on a 624-acre plot of land that lay between the small towns of Balsam Grove and Rosman in rural western North Carolina.

The quickening pace of Soviet space launches in the late 1950s and early 1960s proved the United States needed to gain ground in space launches and tracking technology, not just to win the propaganda war but to secure our nation's defense.[7] Within the United States, the military dominated the early development of boosters, satellites and tracking networks, often overshadowing the interests and needs of civilian space scientists.[8] This convinced the Eisenhower administration that advances to both civilian and military space programs would accelerate under separate management.

OPERATION MOONWATCH: THE ART OF TRACKING MAN-MADE SATELLITES

In the mid- to late 1950s, professional astronomers in the United States weren't sure where the first foreign satellites might appear on the horizon, so they enlisted the support of amateur astronomers. Under Operation Moonwatch, which was initiated in 1956, eight thousand civilians were organized into 250 worldwide teams. They sat arrayed in a line, scanned the nighttime sky and reported their observations using optical telescopes. Until professionally manned optical tracking stations came online in 1958, Moonwatch participants played an important role in spotting the world's satellites. Their information often rivaled the data that was obtained by military tracking stations.

Operation Moonwatch team, Pretoria, South Africa. *Courtesy of the Smithsonian Institution, Image no. 96-960.*

On October 1, 1958, just days before the first anniversary of the *Sputnik-I* launch, President Eisenhower announced the creation of the civilian National Aeronautics and Space Administration (NASA) to spur innovation. The U.S. military and the civilian CIA would focus on developing space technology for national defense, including missiles, boosters and military and reconnaissance satellites.[9] NASA would launch scientific satellites for discovery and civilian applications, and it would have a second mission: manned space flight.[10] In dividing the space pie, the military would give NASA its scientific satellite program, including its twelve-station Minitrack space tracking network and management of the troubled Vanguard program. In return, NASA, by charter, was required to share its "discoveries that have military significance."[11]

NEW TRACKING SYSTEMS NEEDED FOR AN "EXPLOSION OF SCIENTIFIC KNOWLEDGE"

After President Kennedy announced his goal of placing a U.S. astronaut on the moon by the end of the 1960s, NASA's budget soared from $400 million in 1960 to almost $6 billion by 1966.[12] As one NASA manager put it, the manned Apollo program was "the rising tide that lifted all boats."[13]

Funding increases fueled new scientific satellite programs designed to explore the science of space and establish test platforms for communications and other technologies that were useful for the manned program and the commercial sector. To achieve these goals, new Earth-orbiting satellites would travel more deeply into space using paths never flown before: highly elliptical orbits, some over 100,000 miles above the Earth's surface, and geosynchronous orbits, which, at an altitude of 22,236 miles, allowed satellites to hover at fixed points above the Earth.[14]

NASA's hand-me-down space tracking network from the military, Minitrack, would not keep pace with the space agency's dramatic new plans. The three successful Vanguard satellites Minitrack followed in 1958 and 1959 were small and achieved simple orbits and collected a limited amount of data.[15] Minitrack's increasingly obsolete network of tracking stations were not sensitive enough to track satellites in geosynchronous orbits tens of thousands of miles above the Earth, much less scientific satellites in elliptical orbits over one hundred thousand miles in space.[16]

But it wasn't just the orbital distances that would quickly overwhelm old tracking and data collection systems. As NASA scientists gained experience

In May 1961, President Kennedy challenged NASA to land an astronaut on the moon and return him safely to Earth by the end of the decade. *Courtesy of NASA.*

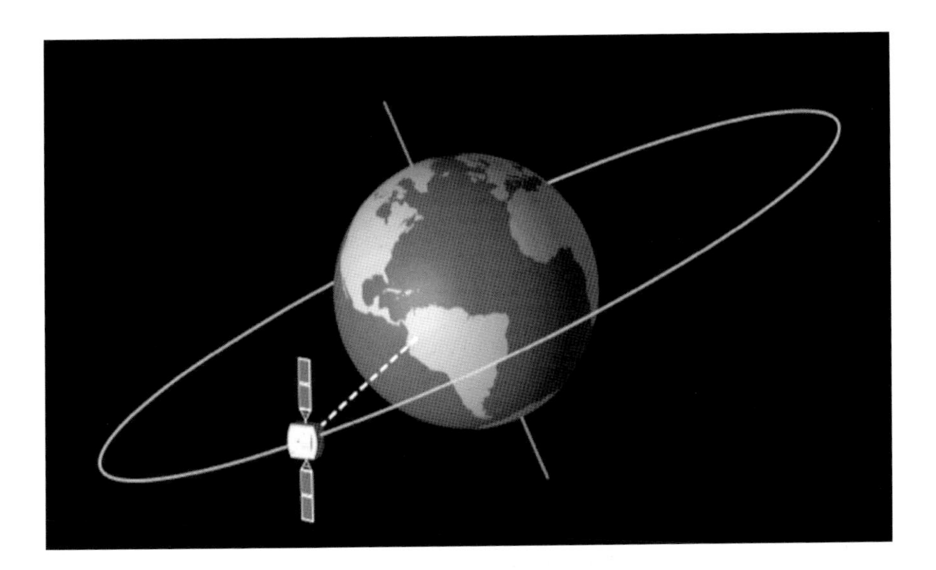

Above: A geosynchronous satellite travels at an angle to the equator at the speed of the Earth's rotation and passes the same fixed point above the Earth at the same time each day. A geostationary satellite stays above the Earth's equator and travels at the same speed as the Earth's rotation so it stays constantly above a fixed point. *Courtesy of NASA.*

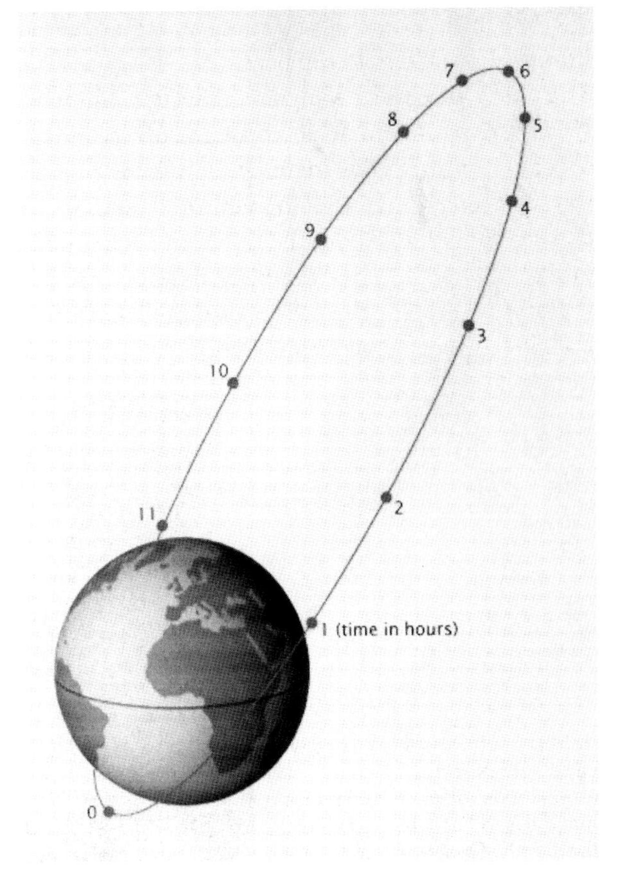

Right: Satellites in elliptical orbits can reach altitudes of over one hundred thousand miles. A Russian Molniya communications satellite orbit (pictured) achieves apogee twenty-four thousand miles above the Earth. The satellite communicates most effectively during hours three to nine, before it picks up speed to complete its orbit. *Courtesy of NASA.*

and technology improved, the number of successful satellite launches and satellite lifespans increased. Vanguard, for which Minitrack had been designed, had an on-orbit time measured in weeks. New improvements in fuel chemistry and engine design would allow satellites to stay productive for a half dozen years or more.[17] With increasing satellite lifespans and with more satellites of greater capability now in space, the burden to monitor and collect data would increase exponentially. Instead of tracking and collecting data from just one satellite, by the mid-1960s, ground stations would perform these tasks for up to forty scientific satellites each day, with one satellite regularly within its field of view.[18]

NASA anticipated the need for a modern network of ground stations, and between 1958 and 1962, Congress authorized over $1 billion for that purpose.[19] New ground stations would be equipped with antennas capable of commanding, tracking and collecting data from scientific satellites transmitting an unrelenting flow of TV, optical, weather, navigation and communications information, beamed to Earth from great distances. Processing the data and separating the important from the unimportant would occur first at these ground stations, where the signals were collected.

NASA planners were right. By the end of the 1960s, the volume of data collected from satellites would explode to about two hundred times that obtained in the early 1960s, representing three hundred hours of space data every day. Every week, each ground station, working around the clock, would collect and store the data on about fifty miles' worth of magnetic tape, which would then be shipped to Greenbelt, Maryland, for processing at the Goddard Space Flight Center (Goddard SFC), NASA's headquarters for unmanned scientific satellites.[20]

ROSMAN: NASA'S FIRST MODERN GROUND STATION FOR SCIENTIFIC SATELLITES IN THE CONTINENTAL UNITED STATES

In 1961, the search was on to find the right spot for NASA's newest and most advanced ground station for scientific satellites. The unidentified site, "Project 3379," was destined to be the key player in an evolving complex of advanced ground stations NASA called STADAN, its Satellite Tracking and Data Acquisition Network.[21]

NASA had exacting specifications for the location of its newest ground station. Most importantly, the site had to be remote, away from power

James E. Webb, NASA administrator (1961–68). *Courtesy of NASA.*

lines, substations, radio towers and airplane routes that might be a source of radio waves that could interfere with the station's communications equipment.[22] Satellite transmissions were of such a low power that today's microwave oven heating a TV dinner a mile away could interfere with the signals coming from the satellites orbiting high aloft. Even vehicle noise from trucks and cars could impact signal quality.[23]

There's no clear record of how the land along the slopes of the Tennessee Bald in western North Carolina's Pisgah National Forest came to the attention of site planners, but it may be more than a coincidence that NASA's then–top official, Administrator James E. Webb, was a resident of rural Granville County, North Carolina.[24]

Regardless of how the parcel was identified, the site just outside of Balsam Grove was added to the list of fifteen locations under review by the site selection committee. NASA named western North Carolina's candidate Rosman because, even though it was located seven miles from the town, it was the closest incorporated municipality likely to be recorded on a 1960s map.

Rosman seemed a good location, logistically, for NASA. It was one of the few areas east of the Mississippi River without regular airplane overflights, yet it was close enough to Asheville's airport for a quick plane ride north to the Goddard SFC in Greenbelt, Maryland, and south, to the East Coast Test Range at Cape Canaveral, Florida, where most of the scientific satellites would be launched.[25]

The Rosman site also had the right geology. The land, deep in the Pisgah Forest, was ideally situated within an ancient seabed that was created 450 million years ago, long before the world's continents drifted apart. Volcanic granite was pushed upward through the ocean's sedimentary rock, and over the millennia, the seabed eroded at a faster rate than the granite, creating a bowl-shaped depression. Nature offered a near-perfect environment for satellite reception, as the surrounding granite walls blocked radio and other electromagnetic waves from interfering with site operations.[26]

Aside from logistics and geology, perhaps the greatest benefit of Rosman was its geography, its latitude and longitude, which gave the site access to the

LUMBER MILLS AND TANNERIES: THE ORIGIN OF NASA LAND

Much of the original 638-acre NASA tract was owned by George Vanderbilt, the owner of the Biltmore estate and grandson of the American railroad and steamboat shipping magnate Cornelius Vanderbilt. After George's death in 1914, his wife, Edith, sold land to the U.S. government and also deeded a portion of 20,000 acres of the Gloucester Range to Joseph Silversteen of Brevard for his Gloucester Lumber, Rosman Tanning Extract and Toxaway Tanning Companies. Aside from harvesting the wood, Silversteen extracted the tannins from oak and chestnut bark. Silversteen's seven hundred tannery workers processed up to five hundred hides per day.

Businessman Joseph Silversteen. *Courtesy of the Roswell Bosse North Carolina Room.*

In 1925, Silversteen sold an option to purchase 10,211 acres of this property to the U.S. Department of Agriculture at $3.75 per acre, or $38,291 for the full tract. The agreement allowed him to continue taking lumber and minerals from the land for up to twenty years. Eventually, the Department of Agriculture purchased the land under the 1911

Silversteen acquired much of the land that became NASA's Rosman site. *Courtesy of the Roswell Bosse North Carolina Room, Transylvania County Library.*

Weeks Forestry Law, which was used to create the first national forest in the eastern United States. The Silversteen acreage was added to land that had already been purchased from the Vanderbilt estate for the newly established Pisgah National Forest.

The creation of artificial tannins in World War II eventually led to the demise of Silversteen's Rosman Tanning Extract Company, and a year after he died in 1958, his Toxaway Tanning Company closed its doors. What remained of the Gloucester Lumber Company merged with Champion Paper in 1967.*

* The Gloucester Range is located along the north fork of the French Broad River, south of Balsam Grove. Joseph Silversteen cut what is now Route 215 to access this timber for his Gloucester Lumber Company. The site NASA eventually selected is in an area of Gloucester Township called the Silversteen Community (named after Joseph Silversteen). Gloucester Township also contains the unincorporated municipality of Balsam Grove. See U.S. attorney general Jno Sargent to U.S. secretary of agriculture William Jardine, April 9, 1925, TLS (photocopy), from Attorney Donald Jordan, Brevard, NC; Scott Shumate, Biltmore House historian, email to the author, "A Question About Vanderbilt Land Holdings," TL, April 22, 2022, Asheville, NC.

widest range of satellite orbits for NASA's nascent satellite programs. Indeed, NASA was looking for an East Coast location that would serve as a hub for tracking and collecting data from scientific satellites in polar, elliptical and geosynchronous orbits. Rosman was well situated, geographically, as it could link up with NASA's two other new tracking and data collection sites that also were equipped with eighty-five-foot-wide dish antennas. These new facilities were located in Fairbanks, Alaska (which would access satellites in north–south polar orbits), and Orroral Valley, Australia (which tracked and collected data from satellites inclined between the poles and the equator). In this way, Rosman would serve "double duty" for north–south and east–west satellite coverage and data reception.[27] A bonus for cost-conscious NASA administrators was the price. NASA would pay nothing for the site, which already belonged to the U.S. Forest Service.[28]

After assessing the pros and cons of each location, the site selection committee whittled the fifteen competitors down to two: Rosman and Fort Valley, located just south of Macon, Georgia.[29] NASA originally favored the Georgia site for many of the same reasons it was attracted to Rosman, but the fact that NASA could get land in the Pisgah National Forest for free ultimately tipped the scales in favor of Rosman.[30]

A "FABULOUS FACILITY" TAKES SHAPE

The Rosman facility is to become the hub of NASA's east–west and north–south tracking...in the United States and abroad for ground support of a new generation of scientific satellites
—NASA press release[31]

A wooded 638-acre parcel was surveyed, and on June 6, 1961, less than three months after Goddard Space Flight Center was dedicated and just three years after NASA itself was established, a memorandum of agreement was signed, allowing NASA to gain control of the Forest Service land. NASA made it public on December 18, 1961: Rosman would host the most advanced scientific satellite tracking and data collection facility of its day, and NASA was in a hurry to get construction underway.[32]

Plans to develop the site took shape, and in February 1962, just two months after its December announcement, ground was broken on the Rosman Space Tracking and Data Acquisition (STADAN) facility. Even before construction was completed, the Rosman site was hailed by the Associated Press as "one of the fabulous facilities of this space age."[33]

NASA had big plans for the Rosman station. It spent over $5 million to develop, build and equip the facility that required an army of electricians and mechanics; engineers skilled in hydraulics, communications, antenna design and operation; and radio frequency and recording technicians, among other specialists. Over two dozen buildings were constructed to house computers for satellite control, offices, shops for spare parts and even a wastewater treatment facility and a pumphouse with two forty-five-thousand-gallon water reservoirs for heating and cooling. Electric generators, using about one thousand gallons of fuel per day (when needed), were acquired to ensure uninterrupted power.[34] A one-thousand-foot-long underground tunnel was constructed for the main power and communications cables to run the enormous dish antenna. More power lines would be installed, crisscrossing the site like a giant tic-tac-toe board. The North Carolina State Roads Commission hired local

NASA's newest ground station located near Balsam Grove was an early member of the space agency's STADAN network. *Courtesy of PARI.*

workmen to improve the existing roads, including two and a half miles of dirt and gravel roads leading to the site, to handle heavy construction equipment. To improve access, the commission also paved Route 215, a winding gravel road from U.S.-64 to the Blue Ridge Parkway at Beech Gap. Employees would later complain bitterly that the zigzagging road, built without guardrails high above the banks of the French Broad River, was a safety hazard during the icy winter months.[35]

• • •

In an interview, Rosman site third-generation employee Brad McCall recounted how his family helped clear the land, which included getting rid of illegal moonshine stills:

> *Just before NASA came, there were three main occupations in Transylvania County: farming, lumber and white liquor. As lumber declined, more illegal stills sprang up, and when NASA settled on the site outside of Balsam Grove, my grandfather Fleming "Jahu" McCall set out to survey the land and remove the stills. He found three—two were abandoned, but the last was*

Clearing the land was a family affair. The McCalls prepared the site for NASA's construction. *Courtesy of the National Archives.*

guarded by a man with a shotgun. Jahu approached the man and told him about the government's plans, but he was reluctant to move. My grandfather became more forceful. "NASA is coming, and you will be removed!" That caught the still owner's attention, and he finally abandoned the site, which was on a little stream located behind what became NASA's cafeteria.[36]

After NASA selected the property for its newest ground station, local workmen from Balsam Grove, Rosman and Brevard were hired to clear the land. Road clearing and grading was a family operation. My grandfather Jahu and my dad, Thad, then eighteen, residents of the local Silversteen Community, were hired to cut the road leading from Macedonia Church Road to the new NASA site—what they called Neil Armstrong Drive and is now PARI Drive. My dad's brother James "Buster" McCall graded the roads in preparation for heavy equipment and construction. Work began on the first big antenna, called 85-I, and operations building before the rest of the land was cleared.

• • •

MASSIVE ANTENNA BECOMES A ROSMAN ICON

The operations building covered with "NASA blue" brick was one of the first structures constructed, as were the transmitter buildings and the powerhouse that stored the generators. Photographs show work began quickly, too, on the signature feature that would forever brand Rosman as NASA's premier scientific satellite space tracking facility of its day: a huge 85-foot-wide dish antenna that would rise over 125 feet in the air and tower over the rest of the site. Though the antenna was fabricated by the Rohr Corporation, skilled workers from Transylvania County put the pieces of Rosman's new icon together. The curved aluminum dish spanned nearly 100 feet, weighed over 325 tons and had support beams embedded in concrete ninety feet underground. It was pricey for its day, costing $760,000 to design, fabricate and erect, and it would consume the same amount of energy it would take to operate two thousand television sets simultaneously. Once finished, NASA gave its brand-new antenna the boring but functional name: 85-I.[37]

Antenna 85-I was the first eighty-five-foot-wide dish constructed in the continental United States specifically for NASA scientific satellites. NASA's plan was to use Rosman's 85-I for observatory-class satellites that would use polar and elliptical orbits and, for the first time, place telescopes in space.

The operations building and the 85-I antenna were the first structures built by NASA. *Courtesy of NASA.*

Contracts for NASA's Rosman and Alaska's Fairbanks eighty-five-foot-wide dishes were negotiated with Rohr Corporation in January 1962, but the Alaska antenna, designed to collect information from satellites with high data rates in polar orbits (including most of the observatory-class), was declared operational in March that year, four months before Rosman's. The Rosman site would have a second eighty-five-foot-wide dish, which was constructed in 1964 so NASA could expand its range of missions to include commanding, tracking and collecting data from a new applications technology satellite (ATS) program that later would test new communications technologies from its perch in geostationary orbit high above the Pacific and Atlantic Oceans.

> *Rosman is the pivotal station whose proximity to Goddard Space Flight Center adds to its importance.*
>
> —*Edmund Buckley, director, NASA Tracking and Data Acquisition*[38]

As NASA's satellites became more sophisticated and its tracks more routine and accurate, satellite tracking receded in importance, and command and data acquisition, collecting the faint signals coming from thousands of miles in space, became the most critical tasks for ground stations like Rosman. And to perform this mission, Rosman's satellite dishes had to be large. Like one might cup a hand behind their ear to hear a faint whisper, eighty-five-foot-wide "big ear" antennas would be configured to hear weak signals, sometimes less than five watts, the power of a child's night light, at distances reaching two hundred thousand miles into space.[39]

The fact that Rosman eventually had two eighty-five-foot-wide dishes reflects the site's preeminent position in NASA's plans. By 1966, the entire global network of seventeen STADAN stations would have just five of these dishes.[40] Moreover, Rosman's big antennas were part of a suite of fourteen new and advanced tracking and collection antennas.[41] Rosman was never retrofit with the increasingly obsolete Minitrack used with first-generation satellites. Instead, it hosted Minitrack's replacement, the new Goddard Range and Range Rate (GRARR) system, which could track satellites with trajectories in a variety of geosynchronous and highly eccentric orbits up to eight hundred thousand miles into space, or three times the distance between the Earth and moon.[42] Rosman also received new command and control antennas and, foreshadowing its preeminent role in an upcoming applications technology satellite (ATS) program, was outfitted with a special antenna suite, which included a direct, high-capacity microwave link to the Goddard SFC.[43]

• • •

The site's iconic landmarks, the twin 85-foot-wide antennas, are monstrously large. Over 120 feet tall and three-hundred-plus tons, these Rosman dishes can withstand winds of over 120 miles per hour because they are stabilized using steel I-beams and iron rebar footers rooted in concrete ninety feet below the ground. Skilled local residents were hired to dig the footers and construct the antennas. Construction accidents were infrequent but did occur; one workman who was on top of a large antenna dish slipped through an open panel and fell ten stories to his death.

James "Buster" McCall, the second of four generations of McCalls employed at the Rosman site, still works there off and on, like he has for the past sixty years, operating heavy equipment. In an interview, Buster told how he and another man had the task of digging ten ninety-foot-deep footers for the two large NASA dishes:

> *We used a three-foot-wide auger to start digging the antenna footers, but as soon as we hit rock, the auger would stop. We'd pull it out, and I'd have to climb into the hole with an air hammer to break up the rock. One man was stationed at the top to make sure there wasn't any trouble.*
>
> *As the three-foot-wide hole got deeper and deeper and as the auger hit more rock, I was lowered by a cable into the hole. I wasn't harnessed, and it was pitch black. I had a lantern on my hard hat so I could see what I was breaking up. The air hammer was loud, and three feet isn't much room. Once the rock was broken, I'd squat down and load rock into a bucket. Then I'd holler up to the guys at the top, who would pull up the rocks by crane and cable. Even though a steel sleeve was placed around the hole, chunks of dirt and rock fell in on me.*

James "Buster" McCall was lowered into four ninety-foot-deep holes to dig footers for 85-I, depicted here under construction. He later dug six additional footers for 85-II. *Courtesy of PARI.*

Water would seep into the hole and was pumped out at the beginning of the morning shift. One day, I was lowered by cable into a deep hole, but the guys at the top forgot to pump it. I couldn't see much, but as I went down into that dark narrow pit, I felt the water. It was getting deeper. Up to my chest. Well, I started yelling, and they finally got the message, and I was yanked out in time.

Me and another fellow worked an hour at a time, eight- to ten-hour days. They paid well.

• • •

EQUIPMENT IS RUSHED INTO SERVICE

NASA was anxious to make its newest and most advanced ground station available for experimental satellites that were nearing launch and worked quickly to make the Rosman site operational. The *Transylvania Times* reported that, four months after ground was broken, "the entire site would be completed, except for the paving."[44] Though its construction was completed in record time, Rosman's full operational capability seemed to be uneven, lurching forward in fits and starts, reflecting NASA's desire to use the site even before it was fully operational. For example, just as construction began and before a full complement of employees arrived, NASA likely used the site in late 1962 to test new mobile GRARR command and data collection equipment.[45] Its target was the first observatory-class satellite, the Orbiting Solar Observatory (*OSO-1*), launched March 1962 into a circular orbit about three hundred miles above the Earth.

It also appears that Rosman's dedication in October 1963 bore little relationship to the site's true operational capabilities. One month after the ceremony that officially opened the site, Rosman's director M. Gary Dennis conceded as much. Speaking before the Asheville Civitan Club, he labelled the site "not fully operational," even though it had just been given its first official task, tracking and recording data from *Explorer 18*, the interplanetary monitoring platform (IMP).[46] Dennis, a native of Richmond, Virginia, was a 1950 graduate of the College of William and Mary and a trained engineer who built computers in his spare time. In February 1963, he arrived with his wife and three children to become NASA's first director of the new Rosman Space Tracking and Data Acquisition facility. Employees recalled that he was a sharp, demanding boss, kind to the rank-and-file but quick to call senior Radio Corporation of America (RCA) managers on the carpet for shortcomings. During his seven-year tenure, Dennis reportedly fired three such managers and kept new senior contractors in line with the threat, "Next time, it's your turn!"[47]

In September 1963, one month before the dedication ceremony, the first group of sixty-five RCA contractors showed up for duty. Only three personnel on site were NASA employees: the station director, his deputy and the secretary. By the time a second eighty-five-foot-wide, 441-ton steel dish became operational in late 1964, 150 more engineers had arrived to support the ATS program, bringing the total complement to slightly more than 200.[48] Over the years, as NASA tasked Rosman with supporting more satellites, this number would rise again—eventually reaching 260 by the mid-1970s. For nearly two decades, Rosman's contract staff ran the facility, working around the clock in eight hour shifts, twenty-four hours a day, seven days a week. While it might have seemed like they lived at the facility, there was no on-site housing. Many called the city of Brevard home, but for some, life in rural Transylvania County didn't provide enough stimulation, so they pushed into Mills River or made the over-one-hour-long commute to and from Asheville.

• • •

In an interview, Jo Ann Jackson, a spouse of NASA Rosman's second director, James C. "Chuck" Jackson, told how they came to join the tracking station:

> *We were native North Carolinians. Chuck was from Lumberton, and I grew up in rural Cleveland County. After Chuck received a degree in electrical engineering from the Citadel in South Carolina in 1951, he was hired by Philco in Philadelphia, which was producing batteries, transistors, radios and televisions. Chuck didn't like it all that much, so he took a position with Lockheed in Marietta, Georgia. He was there for five years, working on telephone voice scramblers. Marietta didn't feel right, so Chuck left in 1956 and was offered a position with NACA* [the National Advisory Committee for Aeronautics, the predecessor organization to NASA].[49]
>
> *It was an exciting time. We got to know astronauts, including John Glenn, who'd soon circle the Earth. Chuck was asked where he wanted to work: Houston or Goddard Space Flight Centers or the Eastern Test Range at Cape Canaveral. Chuck chose Goddard (which oversaw the scientific*

satellite tracking program). He was on the fast track and eventually would direct the division at NASA responsible for space tracking.

We stayed at Goddard in Greenbelt, Maryland, for eleven years and enjoyed it but wanted our children to grow up in an area like rural North Carolina, not suburban Washington, D.C.

We were at a Christmas party when a colleague asked Chuck, "Hey, guess where they're building a tracking station? Western North Carolina!"

Chuck told NASA he wanted to be assigned there, in Rosman. He knew [because it was out of the mainstream] *it wouldn't be the best career move. Chuck's colleagues told him it would be a career buster. That didn't matter. Going to Rosman would be like going home.*

NASA's James C. "Chuck" Jackson, the director of the Rosman Tracking Station from 1970 to 1981. *Courtesy of Jo Ann Jackson.*

At the time, the station had another director [M. Gary Dennis], *so Chuck had to wait his turn by managing a NASA space parts warehouse in Baltimore. It was a can of worms, but it was worth it when Chuck's wish to become station director at Rosman came true in 1970.*

Chuck Jackson's tenure at Rosman would last ten years. He was there to close the NASA site in January 1981.

• • •

DEDICATION CEREMONY UNDERSCORED NASA'S BIG PLANS FOR ROSMAN

On the afternoon of October 26, 1963, the autumn leaves were in full color, making the dedication of the new $5 million NASA Rosman facility a festive occasion. Even Hurricane Ginny, which was off the coast of the Carolinas, wouldn't cloud this day. To the left of the speakers' podium, the Brevard High School band played the national anthem before more than one thousand North Carolinians and invited guests who were seated in wooden folding chairs facing the site's iconic eighty-five-foot-wide dish antenna. Congressman Roy Taylor served as the master of ceremonies. Reverend Dr.

Dedication of the Rosman Satellite Tracking and Data Acquisition Facility

Rosman, North Carolina

October 26, 1963

The dedication ceremony program for NASA's Rosman Station, October 26, 1963. *Courtesy of the Roswell Bosse North Carolina Room, Transylvania County Library.*

Emmett McLarty, the president of Brevard College, led the invocation and Senators Sam J. Ervin and Everett Jordan, and Governor Terry Sanford were on hand to offer remarks. Goddard Space Flight Center's director Dr. Harry Goett told the crowd, "Man will always be a vital part of [space exploration]—whether he sits in the vehicle, as did John Glenn…or at the control panel of the Rosman antenna." Rosman's mayor Austin Hogsed introduced the site's first station director, M. Gary Dennis, as "the newest Tar Heel," and Goddard engineer Hal Hoff concluded the ninety-minute ceremony by riding a cherry picker up to the face of the big dish, where he explained the antenna's major features.[50]

In promotional material, the Transylvania County Chamber of Commerce proudly declared the county "home of the Rosman Tracking Station," though it admitted, parenthetically, that the site really was located in Balsam Grove.[51] Brochures about NASA's newest ground station were located in chamber offices, and tours were offered from 8:30 a.m. to 4:30 p.m., Monday through Friday.

> *The geographic location together with its* [data collection] *capability give the Rosman Station a capacity for ground support to circular orbits in low inclination to the equator, polar orbits of north–south flight and highly elliptical orbits.*
>
> —*NASA press release*[52]

From the outset, NASA had a bold vision for its scientific satellite programs. It offered its satellites as the platform for experimental hardware designed and created by U.S. and foreign scientists in academia, government and commercial enterprises. NASA's only requirement was that the satellite "user" issue a public report detailing the experiment's results. While the

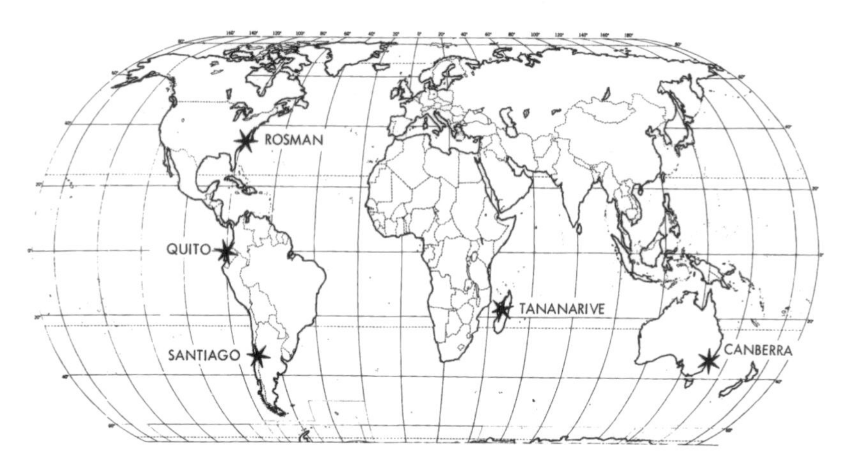

Rosman Station had advanced capabilities and was an important member of NASA's STADAN network of ground stations. *Courtesy of the National Archives.*

Rosman STADAN ground station supported many programs, including, eventually, manned space flights, it is most widely known for its command, tracking and data collection for observatory-class and applications technology satellites (ATS).

Observatory-class satellites, for which Rosman's initial 85-I dish was dedicated, made scientific observations, for the first time, in orbits beyond the interference of the Earth's atmosphere. Three different types of observatory-class satellites (solar, geophysical and astronomical) were placed in various elliptical, polar and low-Earth orbits. A total of nineteen observatory-class satellites were launched between 1962 and 1975. They made discoveries about the sun's influence on the Earth, the magnetic attraction of Earth and moon, the creation and evolution of the universe and other phenomena of near and deep space. This included the first successful space telescope, Orbiting Astronomical Observatory-2, launched in 1968, that spawned a new era of space astronomy that would later include the Kepler, Hubble and Webb space Telescopes.

The ATS satellites commanded from Rosman were trailblazers, too, that enabled breakthrough technologies in weather prediction, cellular communications, global positioning, satellite television and a host of other innovations that we take for granted today.[53]

Aside from these satellite classes, which Rosman is most known for, the site also helped collect data from other experimental craft, including the television infrared observation satellite (TIROS) and Nimbus weather satellites, and Earth resources satellites that used new sensors to peer through the clouds to uncover the Earth's geology, oceans' ecology and forests below. Plans for these experimental meteorological and resource satellites also caught the eye of the U.S. military and intelligence agencies for potential national security uses.[54] When NASA's budget dwindled and its main missions were consolidated in 1972, Rosman played a greater role in supporting communications for the manned space program.

Beyond the hard science and the development of new technologies, few knew these satellites would have the larger impact of sparking the public's imagination. The satellites Rosman commanded took the first pictures of the entire Earth from geostationary orbit, over twenty-two thousand miles in space, pictures that soon would cover newspapers and magazines, inspire the public and create a powerful and unexpected response in some viewers: the photographs of our whole world, in color, established a new awareness of the possibilities for unity and the fragility of our planet.

OBSERVATORY-CLASS SATELLITES: AN EXPLOSION OF SCIENTIFIC KNOWLEDGE

The greatest advance in astronomy since the invention of the telescope.
—*M. Gary Dennis*[55]

Orbiting Solar Observatories (OSO)

Rosman tracked and collected data from eight OSO spacecraft, which allowed the first extended scientific study of the sun. Launched in low-Earth orbits between 1962 and 1975, these OSO spacecraft were used by scientists to add to our understanding of the properties of the sun, including sunspots and solar flares. While these phenomena have been observed since the time of Galileo in 1609, solar research took on greater urgency with the advent of the space age, as new questions were raised about the influence of sunspot radiation on astronauts and spacecraft.[56] OSO spacecraft instruments, which collected data from different wavelengths, including X-rays and visual light, improved our knowledge of the eleven-year solar cycle, when the magnetic field of the sun flips, creating a period of solar turbulence.

One of the first satellites tracked by NASA's Rosman Station, the *orbiting solar observatory-1* (*OSO-1*), launched March 1962. *Courtesy of NASA*.

This is when solar flares are most active and release high levels of radiation that not only endanger astronauts and damage spacecraft but also disrupt radio transmissions from satellites, interfere with space and terrestrial communications and electronic equipment and, if severe enough, could cause power blackouts on Earth.

Orbiting Geophysical Observatories (OGO)

Six OGO satellites were launched between 1964 and 1969, with telescopes focused toward the Earth to study its atmosphere and magnetic field, including the attraction of the moon to the Earth. Three of the satellites, *OGO-1*, *OGO-3* and *OGO-5*, were placed in highly elliptical orbits. *OGO-1* reached apogee at ninety-two thousand miles above the Earth, and because of its distance away from Earth's gravity, its slow speed placed it within

Rosman engineers commanding the first orbiting geophysical observatory (*OGO-1*), which, placed in a highly elliptical polar orbit, was in the site's field of view thirteen hours each day. *Courtesy of the National Archives.*

Rosman's field of view for data collection thirteen hours each day.[57] Three OGO satellites that were launched from Vandenberg Air Force Base in California were placed in polar orbits and studied the upper atmosphere, including the impact of solar rays on the Earth's environment.[58] Most had a design life of one year but remained in orbit for three to five years. All told, the OGO spacecraft sent back to Earth over one million hours of data for scientists to study the relationship between the Earth, moon and sun.[59]

Orbiting Astronomical Observatories (OAO)

NASA launched three OAO satellites between 1966 and 1972 to examine space phenomena, like orbiting bodies composed of dust and ice (comets), rotating stars that emit radio waves (pulsars) and the creation of new stars (novae). The most productive satellite of this class was *OAO-3*, nicknamed *Copernicus*, which, for over eight years, collected X-ray observations of hundreds of stars and long-period pulsars and studied the Earth's atmosphere and the environmental impact caused by the overuse of ozone-depleting refrigerants.[60] The excitement generated by this class of satellites and the trove of information they returned led to the creation of a succession of other space telescopes, including today's James E. Webb Space Telescope. NASA relied heavily on Rosman's capabilities to track and collect data over North America, because the Fairbanks site was out of position to collect OAO orbits. Rosman shared responsibility for supporting these satellites with sister stations in Quito, Ecuador, and Santiago, Chile.[61]

A first day cover issued to commemorate the launch of the *OAO-3* satellite, *Copernicus*, on August 21, 1972. *Author's collection.*

APPLICATIONS TECHNOLOGY SATELLITES: A PREEMINENT ROLE FOR THE ROSMAN TRACKING STATION

Communications satellites can put an end to cultural deprivation caused by geography.
—Arthur C. Clarke[62]

Before satellites, the curvature of the Earth limited the distance Earth-based radio communications could travel, and it wasn't until 1956 that the first transatlantic telephone cable was laid between Canada and Scotland. While scientists understood that using satellites could help overcome terrestrial limitations, the technology had not yet been developed to make instant, reliable and uninterrupted satellite communications a reality. Satellites placed in low orbits, around 200 miles above the Earth, would move constantly and could only see and communicate with about 3 percent of the world at any one time. However, a geostationary satellite placed much farther into space, about 22,250 miles above the Earth's equator, and traveling to match the speed of the Earth's rotation could hover over a fixed spot and communicate continuously with up to 45 percent of the globe. The implications of remaining high aloft over a fixed location were enormous not only for communications but for a wide variety of weather and Earth resource satellites, too. A geostationary satellite was not beholden to the one- or two-hour orbiting schedule for satellites in lower orbits; it could communicate uninterrupted with large sections of the Earth or use cameras to stay focused and see tropical storms brewing days or weeks in advance of landfall. Applications technology satellites (ATS) used new techniques to maintain a stable geostationary orbit while exploring communications signals and developing the technology to exploit them.

The ATS program required new and extensive infrastructure upgrades at Rosman. A second eighty-five-foot-wide dish was erected, and six additional antennas on top of Rosman's bald knob were added to support the six planned launches of ATS satellites. A new power plant was added with a capacity great enough to run two hundred three-bedroom homes simultaneously, and 150 more scientists and engineers were hired, bringing the total number of workers on site to 260.[63] All of this was done in preparation for the launch of ATS satellites, which began in 1966, to explore new communications signals and test different technologies and hardware for commercial and military applications.

All of this—the equipment, personnel and new support infrastructure—was required because Rosman's role in the ATS program went far beyond tracking and data collection. Rosman assisted with ATS launches and insertions into orbit; it commanded the satellites and provided the first look and interpretation of its data. Rosman's ATS engineers also had a separate reporting chain of command that bypassed the Rosman NASA director and went directly to the Goddard SFC. The military had a specific interest in the ATS program, which might have accounted for these procedural anomalies (see Act II, "Spies," page 63).[64]

ATS satellites were not the first geostationary satellites, but they did successfully pioneer new equipment and techniques that were revolutionary for their day.[65] Placed in orbits above the equator, ATS satellites tested different radio frequencies that were useful for satellite, ground and mobile communications and geolocation and new technologies for meteorology and weather forecasting. They provided an important evolutionary link to the creation of new communications, weather and Earth-sensing technologies that, after being developed and tested, had uses for the commercial and defense sectors.

While three of the six planned ATS satellites experienced either launch or in-orbit equipment failures, the remaining three were very successful, and Rosman played an important role in commanding and collecting

Rosman had primary responsibility for applications technology satellites (ATS), which revolutionized land- and space-based communications. (The *ATS-6* is depicted in this photograph.) *Courtesy of NASA.*

scientific data from each of them. Though these satellites were considered experimental, once their capabilities were demonstrated, they were often pressed into service for important real-world missions, like communications support for manned space flights, emergency communications during natural disasters, or were otherwise offered to other government agencies for their own specific purposes. For example, after experiments had concluded on *ATS-1* and *ATS-3*, their imaging cameras were routinely used by the National Oceanographic and Atmospheric Administration (NOAA) for early warnings of potentially catastrophic weather events.[66]

ATS-1*: Achievements in Global Communications and Meteorology*

Immediately after being launched into geostationary orbit above the Pacific Ocean in December 1966, *ATS-1* used a new spin stabilization technique that allowed it to maintain its orbit so it could achieve a number of significant firsts in global communications and meteorology.

ATS-1 proved that satellites could be used to relay communications over vast distances to locations that had previously been underserved or could not be served by existing terrestrial communications networks. Satellite communications held the promise of reducing isolation, increasing greater global awareness and improving educational opportunities. More specifically, *ATS-1* was used to relay college-level educational programming using color television signals to remote areas in Alaska and the Cook, Mariana and other islands in the Pacific Basin. It achieved another first in the telemedicine field when the satellite was used by a doctor, who, through voice relay, offered instruction to a distant Alaskan village on how to stop a patient from hemorrhaging.[67]

> *My wife was the first person to receive a telephone call over a satellite. She was nine and a half months pregnant when I woke her up with an excited call from the Rosman office using* ATS-1. *"How does it feel to be the first person to receive a satellite call?" I asked. After a pause, she responded in a sleepy voice, "It's 5:00 a.m. for God's sake, and it doesn't feel good."*
>
> *—Joe Collins, lead engineer, ATS program, Rosman*

Developing alternate, secure and highly reliable sources of instantaneous communications, capable of traveling great distances, was critically important for a variety of civil and military command and control applications. At the Rosman site, engineers tested different frequencies for space-based communications using *ATS-1*, including a very high frequency (VHF) voice link between Rosman and mobile platforms, such as airplanes in-flight and, eventually, as a communications and data collection link for several of the Apollo manned space missions.

Throughout the Apollo years, Rosman personnel, employing *ATS-1*, played a key role in developing television communications signals and equipment and also refined the techniques that allowed manned space recovery operations to be viewed live across the world. For example, at the conclusion of the historic *Apollo 11* lunar landing in July 1969, Rosman, through *ATS-1*, connected President Nixon to the aircraft carrier *Hornet* to welcome the astronauts home.[68] Rosman also provided the April 1970 "post splashdown" ship-to-shore link used by the ill-fated *Apollo 13* mission.[69] After *Apollo 14*, this proven technology was transferred to the commercial sector, and television pictures of spacecraft and astronaut recovery were beamed through commercial satellites to television networks.

> ATS-1 *was in competition with INTELSAT over which communications satellite would be the first to launch.*[70] ATS-1 *was supposed to launch before INTELSAT, but when NASA's budget was cut, the* ATS-1 *launch was delayed, and INTELSAT beat us by a few months. It became the world's first communications satellite in geosynchronous orbit. Still, we got a bit of revenge. INTELSAT borrowed some* ATS-1 *technology—which they didn't acknowledge, by the way—and the launch delay allowed us to work more on that technology. Well, INTELSAT was being used for the 1976 Summer Olympic Games in Montreal, Canada, when this technology froze up over Maine. The commercial guys at INTELSAT gave us a frantic call and asked if we had a solution. Our solution was to use* ATS-1 *to rescue INTELSAT's beacon, and they didn't acknowledge that either. Rosman had a lot of experience helping the commercial guys. Earlier, we'd broadcasted some of the 1968 Winter Olympics in Grenoble, France.*
>
> *—Joe Collins*

The earliest U.S. weather satellites offered fuzzy images of thick clusters of clouds over portions of the United States every hour and a half.[71] *ATS-1*, which hovered over twenty-two thousand miles above the Pacific Ocean, had an on-board "spin-scan" camera that offered the promise of broad, regular weather surveillance by taking up to fifty photographs a day.[72]

ATS-1, commanded at Rosman, provided this historic first image of the Earth and the moon. *Courtesy of NASA.*

Perhaps the most enduring legacy of the satellite is its highly publicized "first photographs." While *ATS-1* was the first satellite to take a full black-and-white image of the Earth from geostationary orbit, it is best remembered for another first: an iconic black-and-white image of both the Earth and the moon.[73]

Well into the twentieth century, astronomers lacked complete photographic evidence of the Earth's full shape, but that changed with *ATS-1*. Its photographs, along with those of *ATS-3* became more than images; they changed perspectives by offering an evocative vision of our world—a whole and integrated picture of Earth, one speck in the emptiness of interstellar space. ATS images were appropriated by environmental and other groups, which advocated for global solutions to the world's growing problems that had previously been viewed in isolation.

ATS-3: *More Breakthroughs in Mobile Communications, Geolocation and Weather Forecasting*

ATS-1 was followed, in November 1967, by *ATS-3*—but not before a second stage launch failure cut short the plans for a second ATS satellite in the spring of that year.[74]

This time, Rosman helped position *ATS-3* over the Atlantic Ocean in geostationary orbit, where it built on the successes of earlier ATS missions by testing experimental communications equipment. Using *ATS-3*, Rosman, in the fall of 1967, controlled the first operational ground-to-satellite-to-airplane communications link, which provided highly accurate position information to a Pan American aircraft that was flying from New York to London.[75]

The first full-color photograph of the entire Earth was taken by the *ATS-3* satellite, commanded from NASA's Rosman station, on November 10, 1967. It was on the cover of the fall 1968 edition of the *Whole Earth Catalog*. (See insert for color image.) *Courtesy of NASA.*

Rosman employees also helped boost America's global prestige by directing real-time television transmissions of the *Apollo 11* moon landing to network TV for millions of viewers in Venezuela and other South American countries.[76] Even after the satellite concluded its useful life, it was switched on to relay emergency communications during natural disasters, including the 1980 Mount St. Helens eruption in Washington state and the 1987 Mexico City earthquake that killed ten thousand.

The most publicly recognized achievement of this satellite was in photographic meteorology. *ATS-3* improved on earlier satellites' black-and-white weather images by producing NASA's first color photograph of the entire Earth from geosynchronous orbit, over twenty-two thousand miles above the Earth, which became the iconic cover photograph for the *Whole Earth Catalog*.[77] (See back cover or insert for full color image.) Over the course of its eleven-year life, *ATS-3* took complete weather images of Earth, including ninety thousand square miles of the Atlantic, each day. Like *ATS-1*, which provided weather observations over the Pacific, the experimental *ATS-3* was pressed into service to identify dangerous Atlantic hurricanes approaching the East Coast. Rosman engineers were put to the test less than one month after the *Apollo 11* moon landing, when, in August 1969, Hurricane Camille was brewing in the Gulf.[78] Rosman would be at the forefront of a new chapter in weather forecasting.[79]

ATS-6: *America's Most Advanced Communications Satellite*

One TV receiver can start a social, economic and educational revolution.
—Arthur C. Clarke[80]

After the launch failure of *ATS-4* and an in-orbit malfunction of *ATS-5* in 1968 and 1969, respectively, the last ATS satellite, *ATS-6*, was placed into geostationary orbit in May 1974 over the Galapagos Islands in the Pacific Ocean. The largest, most powerful and sophisticated communications satellite in orbit, *ATS-6* carried more than twenty scientific experiments, including nine transmitters operating on seventeen different frequencies.[81] It achieved many space firsts, including one we use today in our everyday lives: direct broadcast satellite TV.

Rosman directed the transmission of television signals from *ATS-6* to small, inexpensive fiberglass satellite dishes in remote locations, where

HURRICANE CAMILLE: ROSMAN AT THE BEGINNING OF HURRICANE WATCHES

It started as a tropical depression off the coast of Africa. The Weather Bureau called: "Can you take some pictures of this?"* *ATS-3* was over the Atlantic, and, being in geostationary orbit, it could take a big picture of the entire ocean. Other weather satellites were in low-Earth orbit but only had small occasional views.

The storm was called Camille, and as it picked up speed, it rolled into the Caribbean and then Biloxi, Mississippi.

This photograph of hurricane Camille in the Gulf of Mexico was taken from *ATS-3* and collected by NASA's Rosman Station. *Courtesy of NASA.*

It was the largest hurricane in modern history. When they told us Camille was slowing down, we stopped taking pictures. But then it gained speed, and they called again, "It's going up the James River in Mississippi. Put the camera back on!"

The Weather Bureau said we gave them the information they needed to warn coastal towns that twenty-foot waves would be rolling in. Weather forecasting was so new, some residents didn't believe us. They said, "We've weathered these before," so lives were lost. But we saved a lot of people, too.

As Camille was rolling out, Debby was rolling in. Then the Weather Bureau started putting up its own satellites. But Rosman was at the very beginning of hurricane watches.

—Joe Collins

* The Weather Bureau was the forerunner of NOAA, the National Oceanographic and Atmospheric Administration, an agency within the Department of Commerce, which, among other responsibilities, forecasts weather.

mountainous and otherwise difficult terrain made television reception impossible. *ATS-6* was also the first to broadcast educational TV on a regular basis. Programs were received in the Rocky Mountains, Appalachia and in the states of Washington and Alaska, and the satellite demonstrated flexibility by transmitting multiple signals to different geographic areas. Rosman also directed *ATS-6*'s two-way color TV signals from medical teaching facilities in the Denver area to ten Veterans' Administration hospitals in Appalachia, located from Georgia to West Virginia.[82] It's the same technology we use today in Zoom and Skype two-way communications.

In May 1975, engineers in Rosman moved *ATS-6* eastward to a geostationary orbit over Lake Victoria in Kenya, where it broadcast four hours of health and education programming daily to five thousand low-cost television receivers in rural villages and cities throughout India.

ATS-6 also achieved a number of other firsts: it deployed a thirty-foot antenna into space, the largest of its day, and it was the first satellite to relay a modified flight plan to an aircraft while in flight. It performed many other tasks, routinely relaying communications and the positions of ships and aircraft, while its cameras took a complete image of the Earth every twenty-five minutes.[83]

In its later life, *ATS-6* was moved back over the United States where it tracked and relayed data from experimental weather and geodetic satellites, Nimbus-6 and the geodynamics experimental ocean satellite-3 (*GEOS-3*), and the Apollo-Soyuz manned space program. These tracking and relay missions were, perhaps, the least known but most important firsts for *ATS-6*, because they would have a direct impact on the future of the entire NASA scientific satellite ground station network, including Rosman.

Previously, satellite tracking and data relay were the responsibilities of the ground-based STADAN system.[84] Now, NASA was using satellites for tracking and relaying information from other satellites and in the future, would rely less on ground stations to perform this function. After the successful July 1969 *Apollo 11* moon landing, cuts to NASA's budget became more pronounced, and by the early 1970s, the space agency was under tremendous pressure to reduce costs. Eliminating ground stations would save NASA millions in station operations and maintenance expenses. This revolutionary approach, moving space tracking and data collection from ground stations to space satellites, would eventually lead to the consolidation and eventual demise of the NASA's extensive ground station network.

A LATE NIGHT RAID BY THE FBI

The *ATS-5* satellite was supposed to spin clockwise about 22,250 miles above the eastern Pacific Ocean. Due to a malfunction, however, it spun counterclockwise about 800 miles up. Still, a few experiments were salvaged, including an infrared (IR) sensor that was pointed toward Earth to study how the atmosphere interferes with radio signal reception.

> *One week after I turned on the IR experiment and was collecting results, I showed up to work, and all of the IR equipment and my stack of results were gone. All of it vanished. Overnight. I was told the FBI showed up late at night and took it all away. They said the IR sensors had been turned on over some sensitive countries. The United States hadn't admitted publicly yet that we had satellites imaging the Earth in lower orbits, and the government was afraid the news would leak out.*
> *—Joe Collins*

Former Rosman engineer Joe Collins with an *ATS-6* satellite on display at PARI. *Courtesy of PARI.*

In the mid-1970s, NASA had interest in launching a follow-on to *ATS-6*, with an "operational focus" employing U.S. military, specifically air force, payloads. A backup ATS satellite was available, but despite the support of thirteen senators, adequate funding could not be found.[85]

NASA'S NEW MANTRA: "BETTER, FASTER, CHEAPER"

In the years immediately following Neil Armstrong and Buzz Aldrin's lunar landing, NASA struggled to identify a future vision of America in space as compelling as setting foot on the moon. NASA's budget, which had climbed so far so quickly in terms of real dollars, peaked in 1966. By 1975, NASA's funding was one-third of its highest level, representing less than 1 percent of the total federal budget. It seemed America had lost interest in space. The race was over. America had won.

By the early 1970s, U.S. human spaceflight had all but ended. The July 15, 1975 Apollo-Soyuz mission was America's only human space flight mission between the last Skylab launch of May 1973 and the first launch of the space shuttle *Columbia* in April 1981. During this dry spell, NASA had to justify its plans to use its dwindling resources in ways that emphasized the practical uses of space, focusing on communications, meteorology, oceanography and remote sensing of Earth's resources.

In anticipation of continuing budget cuts and its new, narrowing mission, in 1972, NASA also began a downsizing program that merged its manned space flight (MSF) and STADAN programs and eliminated redundant

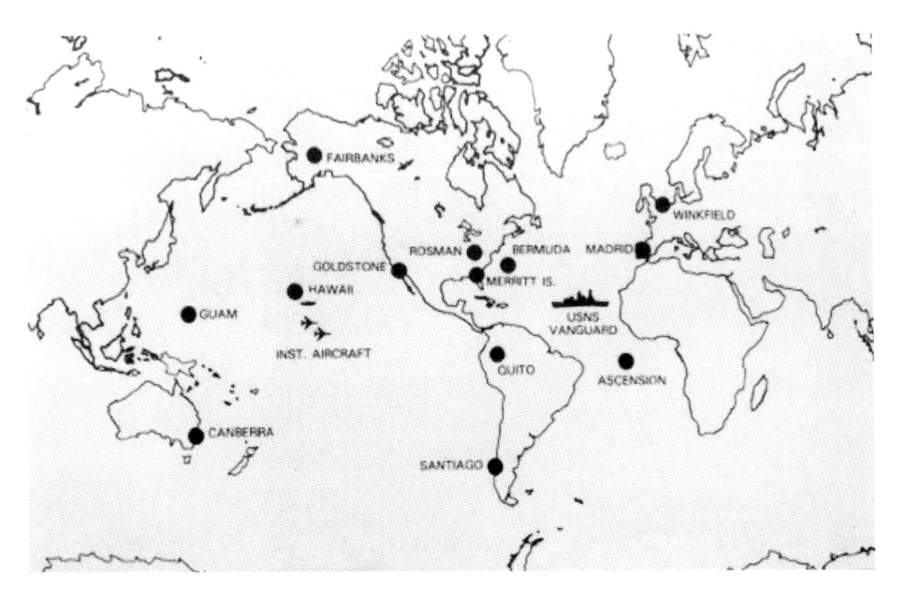

In the early 1970s, budget cuts led to the consolidation of all NASA's ground stations, creating a single Spaceflight Tracking and Data Network (STDN). *From Linda Neuman Ezell's,* NASA Historical Data Book, *vol. 3,* Programs and Projects 1969–78, *408. Courtesy of NASA.*

ROSMAN SUPPORTS A CRITICAL MOMENT FOR THE APOLLO-SOYUZ TEST PROGRAM

In 1972, budget constraints forced NASA to consolidate what were separate scientific satellite and manned spaceflight programs, so Rosman began monitoring communications and tracking manned space vehicles more regularly, including the July 1975 United States–Soviet Apollo-Soyuz test program.* There were tense moments in this, the first international manned spaceflight. In an interview, PARI's chief technology officer Lamar Owen told the story of U.S. astronaut Thomas Stafford's effort to dock the Apollo vehicle that would, for the first time, unite two cold war competitors in space:

> *Docking required clear sight and a steady hand to avoid damaging the vehicles, perhaps irreparably. At a critical moment—and with only ten meters separating the Apollo and Soyuz vehicles—glare*

---

* Political events shaped Rosman's participation in the Apollo-Soyuz test program. In the days leading up to the July 1975 joint United States–Soviet Apollo-Soyuz test project, the first international manned space flight, the leader of the Malagasy Republic was assassinated. Coup leaders demanded NASA close its ground station or pay $10 million in back rent, retroactive to 1963, when NASA first put the ground station in the country. NASA refused, and its Bendix employees were allowed to evacuate. To pick up the slack, *ATS-6*, commanded through Rosman, was given a major new responsibility, providing ground control communications to support the U.S. and Soviet crews.

An artist's rendition of the Apollo-Soyuz space capsules preparing to dock. *Courtesy of NASA.*

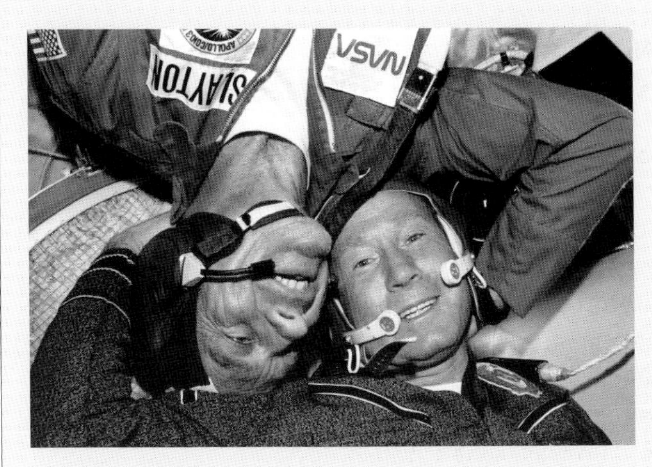

Astronaut Donald "Deke" Slayton embraces cosmonaut Aleksey Leonov aboard the Soyuz capsule. Rosman provided a critical communications link for the space rendezvous. *Courtesy of NASA.*

from the sunlit Earth began interfering with Stafford's view of Soyuz. Stafford later said he was "swearing under his breath," trying to line up the vehicles. Eventually, the Soviet Soyuz vehicle appeared to move toward the Earth's horizon and into view. Just as Stafford got it lined up, Apollo lost radio contact with U.S. ground control.

Minutes later, when Apollo reestablished contact, it was over Rosman, North Carolina. The tense moments had passed. Stafford said they had a real hard time docking, but all hatches were locked and he looked forward to meeting Soyuz.

The good news of the successful docking was relayed directly through Rosman's second eighty-five-foot-wide dish, 85-II.

ground stations. By 1975, nearly half of the twenty-six worldwide MSF and STADAN stations that had begun operations in the 1950s and 1960s were closed. What remained was aggregated into a new, integrated system of ground stations called STDN—the Spaceflight Tracking and Data Network.[86] As part of this consolidation, Goddard SFC assumed more responsibilities as an operating station itself, equipped with state-of-the-art high-speed processing equipment for collecting signals and images from new remote sensing satellites equipped with infrared and other sensors. Goddard updates included advanced data processing equipment, which finally eliminated the need for ground stations to use and transport magnetic tape. By 1978, Goddard SFC also allowed participating scientists to manipulate their satellite experiments directly in real time, instead of working through ground stations. All of these programs, which brought new capabilities to Goddard, were designed to further reduce reliance on the ground station network. But the true death knell for a large global ground station presence was the launch of the highly anticipated tracking and data relay satellite (TDRS).

Based on proven technology from the *ATS-6*, new TDRS satellites, scheduled for launch into geostationary orbits in the late 1970s, would track and collect data from orbiting satellites and relay the information to a handful of ground stations. NASA didn't sugarcoat the impact this would have on current operations: TDRS would "augment and, where practical, replace" ground stations.[87] Millions of dollars in ground operations and maintenance costs would be saved by pushing the tracking and data relay technology into space. There would also be the added benefit of not having to negotiate rent for ground stations or have their status be subject to the political whims of host countries.

The new satellite architecture would be highly efficient. Because of the Earth's curvature, ground stations could monitor only a small portion of a satellite in low-Earth orbit before the horizon would interfere and the satellite would be handed off to another station. A constellation of three TDRS satellites in geosynchronous orbit could provide near continuous coverage of up to fifty satellites in low-Earth orbit, thereby eliminating potential gaps in coverage and ensuring continuous communication with the ground.[88] Moreover, the TDRS system would be given the newest, fastest and greatest capacity collection and relay equipment, increasing the volume of data that could be transferred quickly to one of the few remaining ground stations.

As cost pressures mounted, ground stations struggled to prove their worth. Still, Rosman's future in the early 1970s seemed bright. The observatory and ATS satellites it commanded were still in orbit, and the new geostationary *ATS-6* would be launched in the spring of 1974. But it was the hoped-for new missions that seemed to offer the greatest assurances for Rosman's survival. Its experience with the experimental Nimbus meteorological satellite helped Rosman enter this new age of remote sensing.[89] When the polar-orbiting Nimbus evolved with new sensors, not just to look at clouds but peer beneath them, the platform, in 1972, became the Earth resources technology satellite (ERTS). Equipped with multispectral sensors, ERTS helped to uncover the Earth's riches—mineral deposits, virgin forests and vast ocean resources—while detecting changes to the climate and the environmental damage caused by pollution and ozone depletion.[90] It was a huge new area of exploration with seemingly limitless potential. Rosman supported the ERTS and other

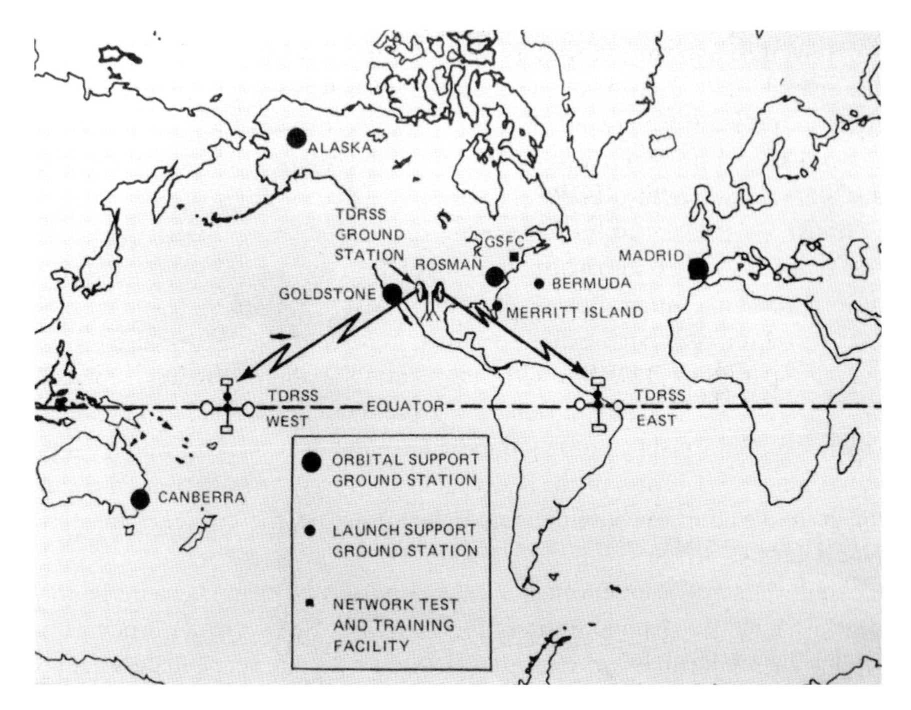

Rosman was expected to participate in the TDRS program (depicted here), but U.S. space shuttle delays led the site to be scratched from the program. *From Linda Neuman Ezell's,* NASA Historical Data Book, *vol. 3,* Programs and Projects 1969–78, *427. Courtesy of NASA.*

science- and imagery-related programs like Skylab by issuing commands that controlled the spacecraft, collecting and directing imagery signals for processing to the Goddard SFC.[91]

But Rosman's future seemed especially promising, because NASA selected it as one of five ground stations to support the new geostationary TDRS satellite program.[92] Director James "Chuck" Jackson reflected this optimism as early as October 1971: "Further expansion [of Rosman Station] will be required within the next two years in order to be ready for new satellite programs around the corner."[93]

NASA's confidence in Rosman's future was reflected in a newly signed thirty-month contract with Bendix Corporation, which would take over operations and maintenance of the site from RCA beginning in January 1973. Jumping on this opportunity in late 1972, Bendix launched an advertising campaign in Transylvania County for eighty new employees. Apparently, the pool of experienced talent was thin, because by mid-1973, the company established an experimental two-month operators training program for local residents with a high school diploma and a willingness to work in shifts.[94]

While Rosman still had missions to perform, what really saved the site from this round of consolidation was the station's aging but still relevant twin eighty-five-foot-wide dishes and GRARR antennas that could collect data from satellites in polar, geosynchronous and elliptical orbits. These capabilities and ground station experience with the ATS program seemed to closely match NASA's evolving mission in the 1970s, which was to apply new communications signals and technologies to overhead platforms for practical purposes.

NASA'S TENURE AT ROSMAN TRACKING STATION COMES TO AN END

The early 1970s were good to Rosman: it had ERTS, Landsat, Nimbus, Skylab, ATS and observatory satellites and even an experiment monitoring the inner ears of frogs in space. But as the decade wore on, the station's bread-and-butter missions began to be whittled away.[95] NASA's cost-cutting strategy of closing ground stations prior to the launch of TDRS placed an unacceptable tracking burden on the remaining stations, which led NASA to begin decommissioning orbiting scientific satellites that had passed their useful life or had successfully transferred their technology to more capable commercial satellites.[96] Rosman's observatory and ATS satellites, some of which had been in orbit for nearly a decade, fell into this category. *ATS-1*, *ATS-3* and *ATS-6* were deactivated in 1978 and 1979. The *ERTS-1/Landsat-1* mission ended in 1978, and by the early 1980s, these remote-sensing satellites would become successful commercial platforms. In the end, only *OAO-3* remained. Rosman's hopes of becoming integral to the TDRS program began to fade when the satellite and the new space shuttle that would launch it became mired in delays. Meanwhile, the Goddard SFC, only four hundred miles from Rosman, was expanding its capabilities, collecting and processing multispectral imagery from Landsat, and was tapped to assume TDRS collection and processing responsibilities. And another new TDRS ground station was being built in White Sands, New Mexico, which became operational in 1983. As the number of satellites passing overhead dwindled, Rosman's management knew the end was near. By the end of 1979, the contractor workforce, once 260 strong, had been cut in half.

While many in Transylvania County believed a bitter pay and benefits dispute between unionized station employees and the Bendix contractor led to Rosman's eventual closure, in actuality, Rosman's demise as a NASA ground station was inevitable and could be traced to years of NASA's budget shortfalls, workforce consolidation and advances in technology that eliminated the need for a large global network of satellite ground stations.[97] The space shuttle *Columbia*'s maiden voyage in 1982 had as a payload the first TDRS satellite. It was the beginning of a new era, in which satellites would take over the tracking and data acquisition support required for NASA's low-Earth-orbiting spacecraft.

NASA made it official in a pre-Christmas, December 7, 1979 press release: Rosman would be part of yet another round of international ground station closures that included stations in Hawaii; Guam; Quito, Ecuador; Santiago, Chile; and Winkfield, England:

> *The Rosman, North Carolina facility expects to cease operations in January 1981. The facility had been supporting the* ATS-6, *which is no longer operating, and the Orbiting Astronomical Observatory, which will complete its mission in November 1980.*

NASA concluded its press release not by consoling its contractor workforce but by expressing pride in its cost-cutting measures:

> *About 5,300 contractor personnel are employed in Goddard tracking and data systems around the world. The effect of these actions will be to reduce this number over the next five years by about 2,300 at space tracking and data network facilities around the world.*[98]

In a December 7, 1979 letter to North Carolina senator Jesse Helms, NASA said it "explored the possibility of the [Rosman] station being taken over by another government agency" but concluded, "there does not appear to be any requirement for a similar operation by anyone. We plan to pursue utilization by others more intensively when a formal public announcement of the station's closure has been made."[99]

A PIONEERING CONTRIBUTION TO GLOBAL SCIENCE AND SPACE TECHNOLOGY

Rosman employees had no reason to hang their heads low. For nearly two decades, Rosman's engineers, technicians and administrators worked around the clock, commanding, tracking and collecting experimental data from

After nearly two decades, NASA departed Rosman in January 1981. *Courtesy of PARI.*

over fifty different manned and scientific spacecraft, monitoring up to forty satellites per day.[100] In doing so, Rosman employees were not just present, but they helped birth our new age of satellite communications, meteorology and remote sensing.

Before the arrival of these scientific satellites, the problems that confronted our country and the world in the 1950s and early 1960s were large indeed. Long-distance personal communications, mostly done through landlines and a transatlantic cable, were slow, costly and of poor quality. Hurricane landfall predictions were rudimentary, which led to loss of life and economic dislocation. We didn't have a good understanding of the Earth's untapped resources, nor did we comprehend the mechanisms by which the Earth was warming with the emission of carbon into the atmosphere. All of these advances and more were the direct results of NASA's early scientific satellites. Spinoffs led to direct broadcast TV, the telemedicine industry, global positioning satellites and cellular and mobile communications. The list goes on. Rosman played an integral part in bringing each of these technical advances to fruition.

Collectively, these achievements improved the lives of Americans and people around the world, they heightened our understanding of the Earth and its environment, boosted America's economic growth, led to our material well-being and made America more secure as a nation. The U.S. scientific satellite program brought us along on an amazing journey of discovery, and the Rosman Satellite Tracking station was a critically important part of it.

Still, NASA's closure of the Rosman station was a tremendous disappointment to the residents of Transylvania County. The space agency offered good jobs that pumped more than $4 million per year into the local economy in the form of workers' wages and expenditures.[101] But the shuttering of the space tracking facility was not the end. It was the end of the beginning of the Rosman site.

ACT II.

SPIES

THE NATIONAL SECURITY AGENCY ASSUMES CONTROL (1981–95)

None of the station's operations are concerned with national defense, and therefore, no classified activities take place at the [NASA] *Rosman station.*
—*NASA*[102]

Rosman: a vital part of the overall security of our country.
—*NBC News*[103]

It might seem that a decades-old ground station cast aside by NASA for newer space-based technology would have little to offer a next tenant. But that would not be true. U.S. intelligence officials saw the site's potential; its antennas, once focused on collecting experimental signals from U.S. communications satellites, could be repurposed to collect the secret communications of foreign adversaries. This insight was both logical and unique. No other ground station is known to have been so transformed from a civilian to a military role. It seems that Rosman, having served the country as a pioneer in the development of space communications, would have a new lease on life as an outpost for the nation's defense. NASA's Rosman Space Tracking and Data Acquisition Facility would become the National Security Agency's (NSA) Rosman Research Station (RRS).

The DoD's National Security Agency took possession of the Rosman site after NASA departed and renamed it the Rosman Research Station (RRS). *Courtesy of NSA.*

THE SEPARATION BETWEEN PEACEFUL AND MILITARY USES OF SPACE: A CONVENIENT FICTION

Rosman, under NASA, was an unclassified facility. Its employees did not possess the security clearances needed to handle classified information, and the site itself was fully open to the public. In fact, brochures about NASA's Rosman station were available at North Carolina roadside welcome centers, and the Brevard Chamber of Commerce enticed tourists to visit the site, where station guards were often enlisted to serve as public tour guides.[104]

Still, NASA's claim that "none of the station's operations are concerned with national defense" was less than truthful. It seems that Rosman and the rest of the STADAN network were no strangers to working with the Department of Defense (DoD). NASA reported that in early 1965, up to 20 percent of the network's activities were in support of tracking DoD satellites, and during the next few years, this level of assistance increased to include tracking and collecting data from air force and army navigation and geodetic satellites.[105] Rosman, like other STADAN ground stations, also filled in where the U.S. military had gaps in coverage and tracked foreign satellites, handing the information to Goddard for forwarding to U.S. Air Force's North American Aerospace Defense Command (NORAD).[106] To maintain security, employees at NASA ground stations like Rosman were not told the origin or function of the military satellites they tracked.[107] Moreover, some of NASA's satellites were dual-use, in that they served civil and military purposes, and Goddard SFC's scientific data was shared freely with the U.S. military, as required by the 1958 Space Act.[108]

Despite the early dependence NASA had on military personnel, hardware and launch facilities, it was quick to cultivate an image of an independent agency dedicated exclusively to the peaceful exploration of space.[109] It was good politics. If foreign governments understood the extent of NASA's ties to U.S. military, it could scuttle the space agency's negotiations to establish its global network of ground stations. Mexico, Zanzibar, Nigeria and Spain prohibited NASA ground stations from supporting military programs. India and Japan opposed classified activities taking place on their soil.[110]

By 1963, NASA had freed itself from a heavy reliance on the DoD's support, but the evolving relationship became more complex and symbiotic. NASA required military hardware, boosters and launch facilities, and the air force needed NASA's test equipment, primarily its wind tunnels. With Goddard's satellite program well under way, its scientific data was especially valuable, as, according to NASA's official history, there were few areas of scientific research without potential military application.[111] The military and the National Reconnaissance Office (NRO), in particular, needed NASA's findings in areas like the composition of the upper atmosphere, including the radiation belts surrounding the Earth; data on the world's magnetic field; new propulsion systems; and experiments using higher-frequency signals for communicating with mobile platforms. This information was useful to the air force's nuclear missile program and the NRO's satellite reconnaissance program, and it enabled the military to develop the equipment it needed to navigate and communicate with its expanding set of mobile platforms: aircraft, ships, submarines and ground forces equipment.[112]

The NASA–U.S. military relationship wasn't always a smooth one. The two sparred about classifying satellite data, especially when it came to encrypting information from U.S. geodetic satellites that, if left unencrypted, could be collected by the Soviets and used to improve the accuracy of their nuclear missiles.[113] NASA refused to classify the information it received, preferring instead to turn sensors off over sensitive countries. The DoD, in turn, complained about NASA's compulsion to publicize its activities, presumably including sensitive military data, to gain public and budgetary support.[114]

ROSMAN PLAYED A KEY ROLE IN COMMANDING ATS SATELLITES THAT HELD MILITARY INTEREST AND PAYLOADS

NASA continued working closely with national security agencies in the 1970s and 1980s to ensure its increasingly sophisticated applications satellite programs met national security as well as civilian requirements.[115]

Rosman engineers were given primary responsibility over the entire series of ATS satellites, and the program was operated independently from the site's STADAN tracking activities. ATS managers reported directly to Goddard, not through Rosman's NASA director, and site engineers were given a great deal of independence, too. They commanded all ATS satellites, inserted them into geostationary orbit and had front-line responsibilities for collecting and interpreting experimental data. Goddard also relied heavily on Rosman to offer advice on how to improve satellite performance.[116]

The military and intelligence communities had a strong interest in the ATS program, and this was a possible reason for the separate chain of command. There was another unusual wrinkle with the program. While NASA accepted experimental payloads from a wide variety of academic, corporate and even foreign governments for the ATS program, all communications experiments were established, conducted and reviewed by the U.S. government, which gave U.S. military and intelligence agencies an inside track on new signal technologies.[117]

The military's interest in the ATS program dated to April 1964, two years before the first programmed launch of *ATS-1*. NASA asked the DoD, through its joint Unmanned Space Panel, for payload recommendations on five future flights of its applications technology satellites. The military took a special interest in communications experiments, including planned capabilities for the deployment of the *ATS-6*'s thirty-foot-wide dish antenna, which unfurled successfully in 1974.[118] The military was probably interested in the cutting-edge *ATS-6*'s millimeter wave and laser communications and geophysical experiments, and it sponsored at least one classified experiment on *ATS-6*, experiment no. 649, which likely originated with the NRO or, specifically, the air force to explore the electromagnetic environment.[119] The planned follow-on to the *ATS-6* was designed to include experimental and operational air force payloads.[120]

Meanwhile, NASA was developing stronger ties with the Central Intelligence Agency (CIA). Associate NASA administrator Robert Seamans Jr. invited the CIA's director of science and technology to a series of comprehensive reviews of the entire NASA program, through which the CIA would gain an intimate knowledge of not only Goddard's advanced technology satellites but its tracking and data acquisition programs, including Rosman's ability to collect and process signals.[121] This relationship included mutual analytic support. In a secret, informal and unwritten program codenamed GALAXY, NASA would help interpret secret intelligence the CIA gathered on the Soviet space program, and in turn, the CIA would provide technical intelligence input to NASA programs.[122] This emerged from an earlier proposal on an expanded NASA contribution to the CIA mission, which included quarterly days-long meetings that would give the CIA routine access to the space agency's information and personnel.[123]

As years passed and NASA's funding dwindled, it made good business sense for the space agency to court the military and civilian intelligence agencies to bolster its budget. The NRO helped NASA by offering contract support, and it shared research and development costs by transferring technology NASA needed for scientific and manned missions. For example, NASA's *ATS-6* satellite, commanded by Rosman engineers, included digital computers used on the air force's *Polaris* submarine launched ballistic missiles, and in a formerly classified program called UPWARD, the NRO provided NASA with high-resolution reconnaissance cameras to map the lunar surface in preparation for the successful 1969 *Apollo 11* mission that landed astronauts on the moon.[124]

The NRO also benefited from the transfer of NASA's technology. In the late 1960s, the NRO found Goddard SFC's information processing equipment, probably used to collect large volumes of weather photographs, "ideally suited to handle current and anticipated needs" for its own photographic satellites. The NRO noted that "Goddard SFC presently supports satellite missions similar to those anticipated by the NRO in the early 1970 timeframe."[125]

It's also likely that some satellite operations commanded by Rosman, ostensibly for civilian purposes, also had intelligence and military applications—and possibly sponsorship. The earlier noted *ATS-6* satellite-to-satellite relay experiment with the Nimbus satellite is a case in point. In the experiment, the *ATS-6* was used for the first time as a relay satellite to track, control on-board equipment and collect and forward information from Nimbus, a low-Earth, polar-orbiting satellite, to a central ground station.[126] While Nimbus was considered a civilian experimental weather satellite, it was originally a joint NASA-DoD project that could have supported the NRO's photographic reconnaissance satellites.[127] In the experiment, the *ATS-6*, functioning as a future TDRS satellite, could have relayed data from Nimbus, standing in for a future low-altitude, polar-orbiting air force weather, communications or CIA photographic reconnaissance satellite. The experiment was, by all accounts, a success. The air force's participation in the *ATS-6* program was acknowledged in the satellite's in-orbit checkout report.[128]

Rosman's role as an East Coast hub for NASA in accessing, processing and interpreting unique communications signals from U.S. satellites in polar, elliptical and geosynchronous orbits could help the DoD perform similar functions against foreign satellites and signals. The fact that Rosman originally was one of five ground stations selected by NASA to support the TDRS relay satellite program suggests a later responsibility for Rosman as a DoD station, possibly in a backup role, commanding TDRS or collecting signals from it, which reportedly supported U.S. military and intelligence reconnaissance satellites.[129]

ROSMAN TRANSITIONS FROM NASA TO THE DEPARTMENT OF DEFENSE

Close coordination allowed the military to have advance notice of NASA's plan to vacate Rosman. By mid-1980, six months before the site was to be shuttered, NASA's site director was invited to the NSA's headquarters in Fort Meade, Maryland, and asked to lead the Rosman transfer to the DoD.[130] NASA's Rosman director would be instrumental in the quick and smooth changeover because of his unparalleled insight into the site's equipment, including the capabilities of its satellite dishes and data processing computers and some of the limiting factors that could impede signal access, like the surrounding topography and potentially balky equipment. He'd also have the opportunity to talk about the NASA workforce and make recommendations for employment with the DoD. After a thorough background investigation and polygraph interview to ensure his trustworthiness, the director received a top secret security clearance, which gave him access to NSA's plans, processing techniques and the frequencies of foreign communications satellites that might be collected from Rosman. No other Rosman employees were asked to join the transition team, though eventually, NSA would hire thirty-four Rosman NASA contractors to work at Rosman Research Station.[131] Rosman's former NASA director remained part of the transition team for nearly eighteen months before returning to NASA in late 1981 to assist with the closing of its Quito, Ecuador ground station.[132]

Security was enhanced when the DoD's National Security Agency (NSA) took over from NASA. *Courtesy of PARI.*

Given Rosman's mission with NASA to track and collect experimental data from U.S. satellites, it seemed natural that the NSA would want Rosman's capabilities to collect, process and report unique communications signals from foreign adversaries.[133] But what signals could Rosman access? During the NASA/DoD transition, the former NASA director at Rosman helped the NSA survey the constellation of ground-, air-, sea- and space-based transmitters and receivers that could be acquired by the site.[134]

TRANSYLVANIA COUNTY: A WELCOMING, COMPASSIONATE COMMUNITY

I was part of the Smokey Mountain Air Force at Rosman. In January 1986, I was in Pete's on Broad Street, wearing civilian clothes, a suit and tie. I was deep in a conversation with another air force officer wearing a suit and tie, too, when a waitress came to us, upset. She said the space shuttle Challenger had just exploded. She must have thought we were NASA employees, that Rosman was still in NASA hands. "I'm sure it's difficult losing your friends," she said, her eyes filling with tears. I still remember the pained expression on her face, like she was informing a family member of a tragedy.

—Fred G.

A helicopter from Fort Bragg had mechanical problems. A bearing in the rotor motor went bad, and it landed in a farmer's pasture. No one was hurt. Rather than be angry, the farmer was proud the 'copter landed in his field. And when people found out that it would take a week to repair and the crew had to stay with the craft, the local folks started feeding them. The airmen ended up giving tours and let the kids sit behind the controls. The misfortune of the 'copter coming down engendered a lot of good will on both sides.

—Art T.

When the DoD officially took over the site in July 1981, it was as if a steel curtain had been pulled across the front gate of the new Rosman Research Station. NASA's transparency, which included daily public tours, evaporated, and physical security was beefed up.[135] "Restricted Area: Unauthorized Entry Prohibited" signs were posted on the chain link fence and nailed to surrounding trees. Pinkerton guards carrying automatic weapons provided 'round-the-clock perimeter protection. Transylvania County was declared "off limits" to Soviet diplomats who might serve as the "eyes and ears" of that nation's intelligence services. Bulletproof glass was installed in the entrance guard post and operations buildings, and a pistol range and helipad were added to the site. Perhaps most telling, a new structure was erected: Building 14, which housed a huge paper shredder. There, red-striped burn bags containing the daily "take" from the site's repurposed satellite dishes were tossed into the machine's steely jaws by a man who was hired specifically because he couldn't read.

A succession of Rosman's chiefs of station, among them, Jim C., Don C. and Jay B., were charged with "maintaining liaison with local civil authorities…to ensure good working relations," but that didn't extend to the local media.[136] Given the secrecy under which they worked, the three chiefs were largely unavailable for interviews with the local newspaper, the *Transylvania Times*. When they did speak, their comments were bland and uninformative. "The mission of the Rosman Research Station," Don once said, "is communications research." When reporters asked about specific targets, including spying on drug cartels or corrupt government officials in South America, the station director waved off the comments as "wild speculation."[137] The chiefs' infrequent public statements were made to allay local concerns and stamp out conspiracy theories that easily took root within the earthy environment of site secrecy. In response to the public's agitation that the military site and, consequently, Transylvania County might be a target for a Soviet nuclear strike, Don told the media, "There are no nuclear materials, weapons or explosives stored or manufactured [at Rosman]."[138] On this point, the station chief was correct, but he didn't answer the question. Yes, there were no nuclear materials or weapons at Rosman, but during wartime, all military facilities, including those that collect intelligence on enemy capabilities and intentions, are legitimate targets.

LOW COST, HIGH CAPABILITIES AND NATIONAL SECURITY CONCERNS DROVE THE NSA'S SELECTION OF ROSMAN

From the military's perspective, picking up Rosman from NASA was a no-brainer. The 1970s were a time of increased tension with the Soviet Union, and a new cold war was brewing. The Soviets were flexing their muscles, developing military hardware to include new mobile nuclear missiles and were expanding their support to Marxist-inspired "wars of national liberation." U.S. policymakers needed more information from the intelligence community, including the NSA, on the Soviets' intentions and plans.

MYTHS ABOUT THE ROSMAN SITE

The presence of large antennas and the secretive nature of the DoD's presence often led to wild speculation about what was taking place behind the Rosman site's chain link fence. Some of the myths might have originated with uncleared local contractors who helped the DoD expand the site in the mid-1980s. Like a game of telephone, each retelling of the work undertaken within the secret facility became more fantastic and eventually became "fact." Third-generation Rosman employee Brad McCall recalled some of these tall tales:

> *When the facility was being built, underground tunnels were excavated to connect power and communications cables from the operations buildings to the antennas. People who saw the tunnels didn't know what they were used for. They were long and narrow, so the story developed that the tunnels held nuclear missiles. Others thought the tunnel was a secret entrance that went from PARI all the way to the Blue Ridge Parkway. One time, when I gave a tour and we entered the tunnel, I heard someone say, "This ain't nothin'. The real tunnel is beneath our feet!"*
>
> *Water was out in an operations building, and I thought it might be a water main rupture. I started digging around looking for the leak. At the same time, there was a dog loose on the compound. I tracked him down and leashed him next to the hole I was digging to find the water line break. Well, someone saw the dog tied up next to the hole, and the next thing I heard in town was that PARI*

There are no nuclear missiles in the station's underground tunnels, just miles of cable. *Courtesy of PARI.*

The PARI staff brings out the plush dolls when visitors ask, "Where are the space aliens?" *Courtesy of PARI.*

was involved in satanic rituals and animal sacrifice.

Some people come to PARI and believe Rosman and Roswell, New Mexico, are related. They ask, "Where are the space aliens?" Well, we have a little space alien dolls we bring out to show them so they're not disappointed.

There are other myths, too:

During the DoD years, the berms toward north ridge were used by snipers. Actually, they're used to control erosion. The site was a Department of Defense submarine base. It was used as a NAZI prisoner of war camp in World War II. There's an underground city beneath the facility. Pisgah Forest controls the weather in the Pisgah Crater, an extinct volcano in the Mojave Desert. Somehow the myths go on and on.

Rosman's huge dish antennas and information processing equipment could contribute to the DoD's enlarged mission to intercept Soviet satellite communications, and the dishes could also be turned to other land-, sea-, satellite- and air-based transmitters throughout the region.[139] Perhaps the best part of the acquisition was the cost. In 1979, NASA planned to give its now-220-acre site back to the U.S. Forest Service. So, when the DoD/NSA took over, it paid nothing for NASA's castoff but still relevant capabilities.[140]

The late 1970s were punctuated by a series of troubling revelations in Latin America. The biggest surprise was the NSA's report of a Soviet combat brigade of two thousand troops and six thousand to eight thousand Soviet technicians in Cuba, supporting the Castro government and manning the Soviets' most sophisticated signals intelligence (SIGINT) site outside of the USSR. The site, called Lourdes, located on the outskirts of Havana, had its antennas trained on the United States. To counter future surprises, CIA director Stansfield Turner, in 1980, listed increased coverage of Cuba as one of the top-ten areas of future intelligence collection and analytic focus.[141]

Above: RRS had at least twenty-three major antennas to collect foreign communications and direct U.S. intelligence satellites. *Courtesy of the DoD.*

Right: A U.S. Air Force Corsair fighter jet over Port Salinas Airfield, Grenada. *Courtesy of the Department of Defense.*

The transfer of Rosman to the DoD in 1981 coincided with both the inauguration of Ronald Reagan as the fortieth president of the United States and a dramatic expansion of the Soviet and Cuban influence in Latin America. By 1981, Soviet military assistance to Cuba had grown to $650 million, a sixfold increase since the beginning of the previous Carter administration. The Soviets had also become the top supplier of military equipment to Marxist rebels in El Salvador, and it had sent arms, materiel and one hundred military advisors to Nicaragua to prop up a Marxist government that had seized power fraudulently.

Soviet support allowed Cuban forces to expand their influence more deeply into Central America. Cuban engineers had begun the construction of airfields in Nicaragua and on Grenada, an island about two hundred miles off the coast of Venezuela, leading to what the CIA called "the increasing militarization of Central America."[142] Cuba's construction of a 9,800-foot-long runway in Point Salinas, Grenada, was particularly worrisome, as it could handle heavy Soviet bombers and allow for the refueling of aircraft to

INTELLIGENCE SOURCES, ESPIONAGE AND SECRECY

There are strict laws prohibiting U.S. intelligence agencies, like the CIA, NSA and NGA (the National Geospatial Intelligence Agency) from collecting information on U.S. citizens. The main focus of these agencies and the fifteen other members of the U.S. intelligence community is on "hard target" countries, closed societies where governments tightly control individual freedom and information. These are mostly communist governments (like the former Soviet Union, China, North Korea and Cuba), radical theocratic governments (Iran) and harsh dictatorships (Venezuela). Because there isn't a free press in these countries, the information policymakers need to protect our country isn't readily available, so America uses its intelligence capabilities to acquire this information.

While the capabilities of the U.S. intelligence community are varied, intelligence information mostly comes from the CIA, which has primary responsibility for human intelligence sources (HUMINT), the NSA for signals intelligence (SIGINT) and the NGA for photographic and multisensory imaging intelligence, or what now is called geospatial intelligence (GEOINT).

For the National Security Agency, SIGINT is a broad category of information that includes communications intelligence (COMINT) and electronic signals (ELINT) not used in communications. COMINT is mostly voice and textual communications from the microwave spectrum. ELINT is usually radar transmissions from military equipment or satellites and provides information on the technical characteristics of weapons or space systems. The NSA has many other responsibilities, including ensuring communications security (COMSEC), which includes denying unauthorized individuals and governments from accessing sensitive intelligence community and military communications.*

---

* There is a tension between secrecy and freedom in democratic governments, but intelligence sources are extremely fragile and can easily be disrupted, resulting in a loss of potentially critical information. Often, intelligence judgments on which our government leaders depend to create sound policies are based on single sources of information, perhaps a human agent working in a sensitive North Korean nuclear weapons industry or a cellular telephone used by a talkative Kremlin politician. If these HUMINT or COMINT sources were revealed, they would be silenced immediately, and the source would possibly be killed. The result would be the loss of valuable intelligence information our leaders depend on. The use of intelligence classifications and codeword compartments are based on the source and sensitivity of the information, and they are ways to limit the number of people who have access to the information, thereby protecting the source and reducing the possibility of sensitive intelligence being revealed. There are four levels of classification: unclassified, confidential, secret and top secret, and there are many more codeword subcategories that further limit the distribution of sensitive information.

support Cuba's forty thousand military advisors and troops already stationed in Angola and Ethiopia.[143]

President Reagan ran his political campaign on the theme of countering Soviet and Cuban expansion, and over the course of his presidency, he nearly doubled the DoD's budget from $176 billion in 1981 to $321 billion in 1989. The new president's goals, embodied informally in his Reagan Doctrine, called for aggressively supporting "freedom fighters" in their struggles against Soviet-backed Marxist governments in the third world.

LOURDES: THE USSR'S LARGEST INTELLIGENCE COLLECTION SITE AND LIKELY ROSMAN COLLECTION TARGET

Rosman provided vital security information. It had the capability of monitoring…operations off the coast of Africa and Central America.
—Brigadier General James Yeager[144]

Lourdes, located ninety miles south of Key West and manned by the newly discovered Soviet brigade, was Moscow's largest, most sophisticated signals intelligence (SIGINT) collection and relay ground station outside of the USSR. The U.S. Congress called it the Soviet's "greatest…overseas military intelligence asset," providing, "between 60 and 70 percent of all [Soviet] intelligence data about the United States."[145] Administered by the USSR's military intelligence (GRU) and the predecessor organization to its civilian intelligence organization, the KGB, Lourdes, according to a Defense Intelligence Agency report, could intercept U.S. secret phone and fax transmissions on military operations from the North Atlantic to the Philippines and collect information from U.S. military ground stations as far away as Alaska and West Germany.[146] The Soviets' SIGINT site also intercepted civilian international telecommunications satellites, including communications channels "the (U.S.) defense department rents on [commercial] spacecraft."[147]

Aside from intercepting U.S. military and civilian communications, Lourdes was a critically important hub that channeled these communications and other military information back to Moscow, possibly including data about Soviet and Cuban military successes and failures and future plans for their expansion into the third world.[148] But its potential as a source of information for U.S. intelligence didn't stop there. According to a

An image of the Soviet intelligence collection site in Lourdes, Cuba, taken by a U.S. SR-71 spy aircraft (1981). *Courtesy of the DoD.*

congressional report, intercepting Lourdes's communications could provide information on "Russian spies operating on the American continent."[149]

It's likely that the NSA used the Rosman Research Station to target the Soviets' central communications center at Lourdes to collect and process unique signals of interest to U.S. intelligence.[150] Rosman's acquisition by the NSA occurred at the time when the intelligence community needed more collection on Cuba to support President Reagan's commitment to counter the growing Soviet military presence on the island and Cuban president Fidel Castro's military expansion throughout the region. But it wasn't just the need and timing that supported a Rosman connection to Lourdes. It was Rosman's location and capabilities, acquired rent free, that likely convinced the NSA to take over the site.

The Lourdes site was a top priority for the NSA's collection, as it was only one of two Soviet signals intelligence collection ground stations outside of the Soviet Union's borders.[151] Rosman and Havana are on the same longitude, 82 degrees west of the prime meridian, making Lourdes directly due south of Rosman. Depending on the articulation of the eighty-five-foot-wide antenna, Rosman could be within the footprint of Soviet satellite signals being transmitted to the Lourdes site, 831 miles to the south. Rosman's location and its history of accessing U.S. satellites in polar, elliptical and geosynchronous orbits suggests a mission to collect Soviet communications from satellites in these orbits in both the Western and far Eastern Hemispheres. The Rosman Research Station could also help the NSA direct its own satellites in geosynchronous orbit over these areas and intercept communications from a variety of Soviet and non-Soviet ground, sea, air and satellite platforms.

MOLNIYA SATELLITES: A KEY MOSCOW-HAVANA COMMUNICATIONS LINK

Cuba's proximity to the United States allowed the NSA to employ multiple "close-in" platforms to collect signals from Lourdes, including those from an array of ground stations, aircraft, satellites, ships and even a series of tethered balloons off the Florida Keys.[152] The volume of high priority communications signals, both open and encrypted, probably required a network of these collectors.

Because of geography and distance, Moscow initially had few options to communicate securely with Lourdes. The northern latitude of the Soviet Union made the launch of geosynchronous communications satellites over the equator a difficult technical challenge that it could not solve until the late 1970s. During this interim period, the Soviets developed Molniya ("lightning"), a satellite deployed to a highly elliptical, twelve-hour semi-synchronous polar orbit. At apogee, the Molniya was at its slowest speed and hovered above the northern hemisphere, allowing at least six hours of uninterrupted communications before picking up speed and losing communications on its descending orbit. CIA analysts reported that Moscow communicated with Lourdes through its *Molniya-1* and *Molniya-3* satellites.[153]

Rosman possibly collected signals from the *Molniya-1* and *Molniya-3*. When the DoD planned to depart Rosman it produced sales brochures that identified the frequency range of both of its eighty-five-foot-wide antennas. Both dishes were tuned to the Molniyas' precise frequency.[154] Moreover, the site's antennas, being on the same longitude as Lourdes, gave Rosman the advantage of having the same look angle for collecting the satellites' downlink as Lourdes. The DoD also had a series of four thirty-seven-foot-wide dishes at Rosman that, if focused to the east, could have captured signals coming from Shchelkovo, a Molniya communications relay station northeast of Moscow that the CIA associated with Lourdes.[155]

A Soviet stamp honoring the *Molniya-1* satellite, which was launched in elliptical orbits around the Earth (1966). *Courtesy of the Post of the Soviet Union.*

First launched in 1966, the *Molniya-1* required a constellation of four equally time-sequenced satellites, each offering six hours of coverage, to achieve near-continuous communications. It was a workhorse for the Soviets, but the satellite wasn't without its faults that led to reduced reliability. Four times a day, there were breaks in voice communications, some lasting several minutes, as a satellite, in descending orbit, passed its communications links to a different Molniya satellite approaching apogee.[156] Molniya also needed multiple antennas and complex tracking and communications relay equipment to overcome the technical challenges of handing off communications links after the satellite's orbit took it below the horizon.

ROSMAN ALSO LIKELY ACCESSED GEOSTATIONARY COMMUNICATIONS SATELLITES FOR INTELLIGENCE COLLECTION

As Rosman was being turned over to the NSA, the Soviets overcame the technical difficulties of placing geostationary satellites in orbit and began launching Raduga ("rainbow") and Gorizont ("horizon") as part of its Statsionar communications program.[157] Even as the Soviets continued deploying Molniya, beginning in the early 1980s, Statsionar's geostationary satellites were being launched with greater frequency. Since only one geostationary satellite was needed for uninterrupted communications, Raduga and Gorizont probably began taking over Molniya's high-priority, time-sensitive military communications between Moscow and its forces deployed around the world, including Cuba.[158] By virtue of its capabilities and location, the Rosman Research Station was capable of reaching both these geosynchronous satellites deployed over the Atlantic Ocean.[159] Rosman's ability to access Gorizont was not a stretch, as CIA analysts in McLean, Virginia, five hundred miles north of Rosman, began receiving unclassified Soviet news that was being broadcast from the satellite with dishes installed at the CIA's headquarters in 1985.[160]

The Soviets' Raduga geostationary communications satellite off the coast of South America would have been an easier collection target for Rosman than the Gorizont, which was typically positioned close to the coast of west Africa. Like it had with the Molniya satellites, the DoD sales brochure identified the collection frequencies of Rosman's largest dishes, which were nearly identical to the transmission frequencies of the Raduga and Gorizont satellites. One of its eighty-five-foot-wide dishes could capture the downlink of Raduga's "X band" military frequency.[161] The other eighty-five-foot-wide and one forty-foot-wide antenna were tuned at the same downlink frequencies as the Gorizont's "C band" and "KU band."[162] All three satellites, the Molniya, Raduga and Gorizont had commercial voice or TV channels and also were capable of sending encrypted military communications.

Rosman could also access U.S. military and intelligence satellites in low-Earth orbit over the United States' East Coast or in geosynchronous orbits over the Atlantic and eastern Pacific Oceans, including the channels the U.S. military leased from commercial geosynchronous satellites, like TDRS.[163] Since Rosman, under NASA, was one of the five sites scheduled to serve as a TDRS ground station before the facility was closed, it must have had exceptionally good access to some TDRS satellites and possibly possessed processing equipment that was needed to collect and relay TDRS data.[164] If Rosman had a mission to support U.S. tactical military operations in the Caribbean, like the October 1983 U.S. incursion into Grenada, it could have done so by communicating through TDRS or other military or intelligence satellites in the region.[165]

THE DOD INVESTS HUNDREDS OF MILLIONS OF DOLLARS INTO THE ROSMAN STATION'S INFRASTRUCTURE IN EXPECTATION OF FUTURE SUCCESS

The NSA took its time in evaluating the Rosman site's value to the overall national security effort, and at first, it did little to upgrade the facility after arriving, unofficially, in January 1981. The Rosman workforce prided itself on "making do" with what it had to work with. In fact, DoD employees referred to the Rosman Research Station as "Rosman Radio Shack," a reference to the inexpensive former electronics retailer, where they purchased equipment to keep the site running. "We're trying to take advantage of the capital investment left by NASA," Station Chief Jim C. said. "But the buildings need work."[166] It seems that the Rosman Research Station, valued at a little more than $18 million in 1981, had more than proven its worth and had grown in its responsibilities, and by 1985, the DoD began to invest heavily in the two-decades-old facility.[167] Reportedly, $200 million were spent on capital improvements that included upgrades to nearly every building.[168] A redundant cooling system and a $500,000 electrical substation were added to ensure uninterrupted operations.[169] A new contract with Raytheon was signed in 1986, which allowed the contractor base to grow from 250 to 300.

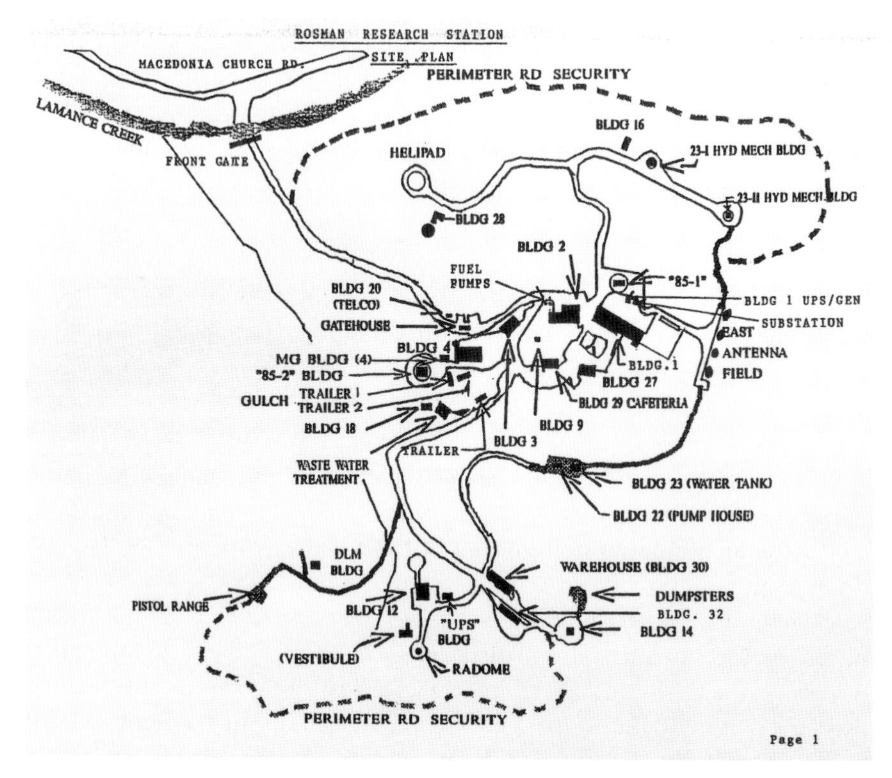

The site plan for NSA's Rosman Research Station. *Courtesy of PARI.*

Eventually, the NSA had a payroll of 450 personnel of all kinds working at Rosman: Bendix, Allied Signal, TRW and IBM contractors; government workers from different agencies; military officers and enlisted personnel all worked shifts, at all hours of the day and night. Satellite dishes sprang up like mushrooms after a warm summer rain.[170]

While the NSA, to this day, remains close-mouthed about Rosman's specific role and successes, the fact remains that the DoD would not have rewarded the facility with a substantial new investment and increase its workforce by more than one-third if the facility had not been successful in intercepting and processing high-value foreign signals of military and intelligence value. The DoD's investment into Rosman not only was a vote of confidence but an expectation of future accomplishment.

Rosman, an NSA field station, probably wasn't a primary site for collecting routine intelligence and military communications coming from Lourdes or other transmitters in the Western Hemisphere; the NSA had a network of highly capable land-, air-, satellite- and sea-based platforms for that task.[171] Rosman's role, more nuanced and potentially more important, was likely related to its earlier NASA mission: to collect, process and interpret unique signals. Most of Rosman's potential DoD missions were aligned with its acknowledged role as a communications "research station."

MANY DISHES, DIFFERENT FREQUENCIES, VARIED MISSIONS

The chief, Rosman Research Station...is responsible for the collection, processing and forwarding of signals of interest to designated recipients.
—NSA Organizational Manual[172]

The Rosman Research Station, under the DoD, eventually had twenty-three major antennas, some with white radome covers that provided environmental protection and kept prying eyes—and possibly Soviet photographic satellites—from identifying the direction of the antennas' targets.[173] The NSA added a third big antenna, similar to the two eighty-five-foot-wide dishes NASA left behind, and had just completed the construction of a fourth, an eighty-two-foot-wide antenna, when it decided to vacate the site in 1995.[174] While the precise number of major and minor antennas located at the site is not fully known, it's likely NSA possessed many more than NASA had when it operated the facility.[175] Moreover, the capabilities of the NSA's antennas were varied and seemed to span the full frequency spectrum. At the low end, wooden poles left in the ground after the NSA departed in 1995 suggest Rosman possessed an extremely low-frequency (ELF) antenna that used ground waves to communicate with submarines. At the top end of the frequency spectrum, the fifteen-foot-wide Smiley dish was used during the NASA years for communications experiments with the *ATS-6* satellite in the extremely high-frequency (EHF) and millimeter wavelengths. EHF signals pack an enormous amount of data but are susceptible to the effects of weather, so they are most useful for satellite-to-satellite communications.[176] The NSA could use ELF and EHF antennas and every dish in between to find unique foreign signals of interest at all ends of the spectrum. As a corollary to this research, Rosman could also help the NSA's defenses by developing more capable, intercept-resistant communications.

A thirty-nine-foot-wide dish enclosed within a radome. *Courtesy of PARI.*

IDENTIFYING "HIGH VALUE" FOREIGN COMMUNICATIONS SIGNALS FROM MOBILE PLATFORMS

The Soviets encrypt their highest value communications to keep them hidden from foreign ears, and this was the case with sensitive military and intelligence communications that were sent to and from the Lourdes site through its Molniya, Raduga and Gorizont satellites.[177] Rosman, under NASA, collected, processed and interpreted a high volume of unique and often experimental U.S. government communications signals. According to the NSA's organizational manual, Rosman had similar responsibilities under the DoD. One of its primary functions was to conduct search operations to identify, classify, process and analyze signals of interest.[178]

Rosman was likely given substantial autonomy to search for high value communications from land-, sea-, air- and satellite-based foreign transmitters to which it had access. Aside from the Soviet Union and Cuba, China, also a player in the Angolan Civil War, could have been a priority target for signals collection and research.[179] After identifying and studying the signals, including the frequency, time and duration of their use and level of encryption—and, if possible, after decoding them—Rosman was charged with distributing the information to other designated recipients. This would always include the NSA's headquarters, especially if further decryption processing was required, and sister ground stations, where the same signal might be collected from other foreign transmitters.

Rosman's research could have also included collecting and processing other high value signals from satellites, such as communications to and from Soviet diplomats and intelligence agents within the United States that were distributed through Lourdes.[180] According to a congressional report, Lourdes received KGB agent communications from a constellation of three Soviet store and dump satellites that operated at low, near-polar circular orbits.[181] The CIA believed the Soviets' new geostationary satellites were best positioned to relay this kind of secret information from Lourdes back to Moscow.[182] In addition to accessing information from geostationary satellites like Raduga and Gorizont, Rosman could have collected signals directly from Soviet photographic and other reconnaissance satellites that were flying low-Earth orbits over the United States; NASA's motors on Rosman's huge eighty-five-foot-wide antennas and new DoD-installed control systems made the twin dishes capable of tracking satellites traveling at speeds of seventeen thousand to eighteen thousand miles per hour.

THE LEGEND OF "SMILEY"

Smiley, the fifteen-foot-wide dish at the entrance of the site is the subject of PARI myth and legend. During its NASA days, the dish, not yet painted with a face, was used to communicate with *ATS-6* and tested signals at super high-frequency and millimeter wavelengths.* Soon after the DoD arrived, a simple painted face appeared on the dish. After the DoD departed, rumors began to circulate. One person said the face at one time sported a large tongue sticking out. Another said it was painted by a disgruntled Bendix employee because he couldn't get spare the parts needed to refurbish the dish. The following story is told by Jim Chester, a former communications technician who worked as a Bendix contractor for the DoD. PARI president Don Cline added his own take based on conversations with a former Rosman NSA chief of station:

It was 1981 or 1982, and the station chief called Rosman's chief of facilities and asked him to paint a smiling face on the fifteen-foot dish. The facilities manager called downtown Brevard for a painter, but no one would do it. So, I said, "I'll paint the face." I got out a sheet of paper, created a pattern, put it on the dish and painted it.

—Jim Chester

I heard that Smiley was painted by a DoD employee without receiving permission from the station chief. The chief ordered it removed, only to have that order countermanded by more senior officers at NSA headquarters who thought the image amusing.

—Don Cline

* This narrow wave has advantages: it contains more data than other signals, but it can be degraded by moisture in the atmosphere. At first, NASA wanted to test these signals in sunny Fort Myers, Florida, but had second thoughts. Wanting to give the signal a more rigorous test, NASA chose Rosman specifically because of its wet weather. To support testing, Rosman employees stomped through the mud to measure the rainfall collected in twenty-four buckets scattered throughout the facility. On at least one occasion, they encountered copperheads and, reportedly, an active whiskey still that was guarded by a man with a shotgun.

Smiley, the fifteen-foot-wide antenna, was one of four antennas left behind by the NSA. *Courtesy of PARI.*

We may never know the complete story, but in either case, it seems that Smiley was painted as greeting for the Soviets who flew their own secret photographic reconnaissance satellites over the Rosman Research Station. It was a way for Rosman's DoD employees to tell the Soviets, "We know you're looking at us—have a nice day!"

> *I heard the Russians took pictures of Smiley....Little did I know the face would still be there more than twenty-five years later.*
>
> *—Jim Chester*

> *Some Russians who visited PARI said they'd seen photos of Smiley taken from their own satellites.*
>
> *—Don Cline*

Above: President Reagan briefing Soviet SS-20 nuclear missile deployments in the eastern USSR, National Press Club, Washington, D.C. (November 18, 1981). *Courtesy of the National Archives.*

Right: An SS-20 intermediate range ballistic missile launch from a road mobile platform. *Courtesy of the DoD.*

Access to the geostationary Raduga and Gorizont satellites also gave Rosman, potentially, a window into other high value Soviet signals. The CIA believed that during the 1980s, the Soviets would use geostationary satellites in areas where Molniya was not feasible for command and control communications with small mobile platforms aboard naval surface ships and submarines, airborne command posts, strategic bombers and, perhaps the highest value target for the United States, Russia's new land-mobile nuclear missile force.[183] This included both the SS-20 intermediate range ballistic missile (IRBM), deployed in 1979, which threatened America's European allies, and the new SS-25 intercontinental ballistic missile (ICBM).[184] With a range of 6,500 miles, the SS-25 could endanger all the major cities and military facilities in the continental United States.

President Reagan realized the difficulty the U.S. military would have in targeting a highly mobile nuclear weapons force, so in the summer of 1985, he issued a national security directive that called on the U.S. military and intelligence agencies to, "on an urgent basis, develop a program…to attack relocatable targets." The plan, which was to be completed on April 2, 1986, included intelligence sensors (antennas, photographic satellites and other capabilities) that were needed to track and target these nuclear weapons.[185] The president's demand was immediate and unforgiving, given that the first SS-25 units were operational. This program to find and target the Soviets' mobile nuclear forces became the highest priority in the intelligence community.[186]

While we do not know if Rosman had specific responsibility to help identify relocatable (mobile) nuclear missiles, the highly encrypted command signals directing these forces probably passed from Moscow through geostationary

A mobile SS-25 intercontinental ballistic missile on static display. *Courtesy of the DoD.*

satellites, some in Rosman's field of view, directly to weapons deployment sites located to the west of Moscow.[187] Moreover, Rosman, under NASA's *ATS-6* program, was familiar with evaluating the unique satellite signals that were transmitted to and from mobile platforms, capabilities that track closely with the immediate demands of President Reagan and the DoD.

ROSMAN RESEARCH STATION LIKELY HELPED TO SECURE AND DEVELOP U.S. SIGNALS

The NSA also has the mission of ensuring that both unclassified and classified U.S. military signals remain secure from unintended recipients. In collecting U.S. military communications from Soviet intercepts that were being relayed to Moscow from Lourdes, Rosman could have helped the NSA understand the state of U.S. communications security (COMSEC), including, perhaps, the Soviets' ability to break encoded U.S. communications. This information would have been useful in patching potential holes in U.S. COMSEC by creating new codes or moving these communications to more secure fiber optic cables.[188] An NSA press officer, in response to a question about Rosman's "communications research" role, said it included intercepting and processing foreign intelligence and "developing codes to protect U.S. communications."[189]

The NSA conducts signals research to improve the security and quality of transmitted data, and it's likely that Rosman performed this kind of work, too, by exploiting new signals for a variety of fixed and mobile platforms. Moving up the communications spectrum to higher frequencies offers a special advantage that would play to the strong research and development base of U.S. defense industries. Generally, the higher the frequency, the greater the complexity required for design and fabrication of electronic components and circuitry. Using complex signals at the top end of the communications spectrum also imposes cost burdens and stresses an adversary's research and development capabilities if it wanted to intercept these new U.S. military signals.

Rosman, under NASA, conducted U.S. government communications experiments and researched, with its *ATS-6* satellite, signals above super high frequencies and into the millimeter range using its fifteen-foot-wide Smiley dish. Rosman conducted laser communications experiments, too. The NSA could have used Smiley or other receivers on site to collect and process these novel signals coming from foreign transmitters.

ROSMAN RESEARCH STATION ALSO USED FOR SPECIAL FORCES TRAINING

Art T. (former army officer): We were told that a special forces team from Fort Bragg would be coming to Rosman to train on us, but we didn't know exactly when. The special forces folks were supposed to enter the facility secretly and place a marker on a satellite dish to prove they successfully breached our defenses.

Well, the day came. They thought it would be a piece of cake. The hicks from the mountains of western North Carolina couldn't be a match for trained special forces officers.

Don C. (former NSA station chief): What they didn't know is on that very day, we'd asked a western North Carolina SWAT team to join our mostly Pinkerton security force. I was walking with my lunch group on Pumphouse Ridge, right by the ninety-thousand-gallon fire suppression tank, when the bushes came alive.

The mountain men from North Carolina successfully sniffed out the special forces operation.

Art T.: I was told it was the only time a team was discovered.

*Don C.: I was on the top of Pumphouse Ridge when they left. Their helicopters flew overhead, and they opened up on me with a machine gun firing dummy rounds. I guess it was their way of waving goodbye.**

* PARI fiftieth anniversary celebration video.

OTHER HIGH-PRIORITY MISSIONS

President Reagan would not allow Soviet and Cuban expansion to go unchecked and was willing to commit U.S. military support, including ground troops, to the Caribbean when necessary. The United States' invasion of Grenada in 1983, its 1989 incursion into Panama to depose Manuel Noriega and its support of the Contra rebels in Nicaragua, the anticommunist military government in El Salvador and, farther afield, the insurgents who opposed the Marxist government in Angola are examples of times during the NSA's tenure at Rosman when the U.S. committed military arms, advisors, troops or paramilitary forces to conflict zones. During military operations that placed American lives at risk, all available intelligence platforms would likely have been used to support U.S. forces. These platforms, serving in a primary or backup capacity, would have alerted military commanders in the field and U.S. policymakers at home about an adversary's intended military actions or political plans. It's likely that Rosman served as an important node of this communications network to support U.S. commanders or offer America's political leaders insight into Soviet thinking. For each of these events, including the 1990 Iraqi invasion of Kuwait and subsequent Gulf War, the Rosman Research Station, with its multiple dish antennas, would provide at least some backup military intelligence support.[190]

Rosman Station likely was involved in supporting U.S. programs in Latin America, including efforts to counter narcotics trafficking. Acquiring intelligence to stem the flow of narcotics from entering the United States became a key intelligence priority during the 1980s and even led First Lady Nancy Reagan to institute a "Just Say No" nationwide campaign to discourage American youth from starting down the path to drug addiction. Indeed, the former president of Panama Manuel Noriega was indicted as a narcotics trafficker tied to Colombian drug cartels. Rosman, as a key regional collection-capable station with access, could have been tasked with identifying communications from foreign satellites used by various drug cartels, narco-politicians and corrupt military leaders in states like Colombia and Panama, to identify the efforts of drug traffickers to penetrate the United States' borders.

WHAT PERSONNEL WOULD ROSMAN NEED FOR THESE MISSIONS?

Rosman was a joint NSA/CSS (Central Security Service) field station, which meant that in addition to NSA civilian officers, cryptologic components of five military services (army, navy, air force, marines and coast guard) could have been present at the site.[191] Given the wide range of possible missions and targets, it's likely that at least the army, navy, air force and coast guard contingents would have complemented NSA civilian personnel on site to conduct their own communications research projects and, working with NSA, distribute findings to their respective commands. The NRO, which had an interest in NASA/Goddard's experimental communications and scientific research program, would have similar interests and possibly participated in Rosman's unique communications research programs.[192] The air force was probably present to help monitor the performance and collection take of U.S. and foreign intelligence satellites and, as with the CIA, would have a presence at the research station through its NRO representative. The CIA would also be available, potentially, as subject matter experts to examine and provide context for Soviet, Cuban, Chinese or Middle Eastern military or political communications intercepts, and crime or narcotics information flowing through the Rosman station. CIA also could assess the performance and direct its own photographic satellites or, in the area of counterintelligence, evaluate collected information on KGB or diplomatic communications from the United States. The coast guard would likely be present, too, if the satellites Rosman commanded collected information useful to keep narcotics from entering U.S. borders.

Collecting, processing and relaying information requires a team of experts with a wide range of skills. Linguists who knew Russian, Spanish, Chinese and Middle Eastern languages would have been present to provide a preliminary real-time read-out of the data collected on site. High value, time-sensitive information would be transmitted immediately to NSA headquarters for further review and rapid dissemination to intelligence, military and policy consumers.

Also, personnel probably would be on hand to evaluate Rosman's collection and perhaps retask U.S. intelligence collection platforms or refine their search methodology used against foreign satellites or other signal emitters to further improve the quality of the intelligence collected.[193] For difficult encrypted signals, signal processing experts would likely try, initially, to get a sense of how the signals were used and their usual patterns and durations so that inconsistencies, like changes in the volume of traffic, might offer clues about evolving states of adversaries' activities, including military readiness.

SECRETS REMAIN: END OF "PHANTOM MISSION" LED TO THE DOD'S DEPARTURE FROM ROSMAN

While speculation about Rosman missions for the DoD are based on the site's activities under NASA, the historical context of Rosman's acquisition, declassified documents and its antennas' known frequencies and "reach," many secrets remain. Why did the NSA cease operations on March 31, 1995, and, six months later, in September 1995, abandon the site to the U.S. Forest Service?

NSA documents say "changing priorities and the need to effect budget cuts" led to the Rosman Research Station's demise as a DoD communications intercept ground station.[194] Three and a half years earlier, in February 1992, President George H.W. Bush and Soviet president Boris Yeltsin met to officially declare an end to the Cold War. Turning down the heat on America's greatest existential threat offered a rare opportunity for Congress to scale back the military budget. This peace dividend took the form of a base realignment and closure (BRAC) program that reviewed military bases to determine their utility and possible termination based on the changing nature of the evolving threat to America.

The BRAC process went through multiple iterations, with commissions established in 1988, 1991, 1993 and 1995. Rosman was not on the list for closure for the first two, and there is no record it was on the 1993 list either. In fact, Rosman's position in the early 1990s seemed secure—so secure that in late 1992, the DoD selected a new contractor for Rosman. Bendix would take over from Raytheon, and the contract period would extend four more years, through September 1996. It also appeared that there would be new missions for Rosman, too. By 1994, a new eighty-two-foot-wide dish antenna, the site's fourth big dish, was under construction.[195]

On June 1, 1993, the NSA's board of directors was presented with a slate of stations that might be placed in caretaker status, or closed and "returned to the [military] service or foreign host."[196] The then–director of NSA, Furman University graduate Admiral J. Michael "Mike" McConnell, would coordinate with Congress to make the final decision on the termination of some sites by the end of fiscal year 1994, taking into account "political and operational concerns."[197] Other NSA documents suggest the decision to close Rosman was made hastily in January 1994 and based on an expectation that the site's last mission would be "shut down in November 1994." Apparently, the NSA wanted a quiet departure, but plans for the site's closure were revealed in the local press before its mission concluded.[198]

For nearly three decades, NSA officials have remained tight-lipped on what that last mission, shut down in late 1994, might have been. Most of Rosman's missions seemed to be ongoing. Signals research is a regular activity without a finite end, and Rosman's set of possible targets had not dried up either. Drugs and narco-politicians continued to ply their trade in

Latin America, and there was no evidence that former Soviet, now Russian, clandestine agents were suspending intelligence collection activities inside the United States. Moreover, the Soviet-era satellites that were available to Rosman, the Molniya, Raduga and Gorizont, were still being launched into orbit regularly.[199] Lourdes, still a valuable target, had begun to decline in value, as U.S. microwave transmissions began to be encrypted in the late 1980s and sent over fiber optic cable with greater frequency. Still, the SIGINT collection and satellite command center at Lourdes remained in operation until 2002. The one area of diminished need could have been the collection of Russian nuclear weapons communications.

The U.S.–USSR Strategic Arms Reduction Talks Treaty (START-I), signed on July 31, 1991, was entered into force on December 5, 1994, around the same time that Rosman closed. The treaty set strict limits on the number of strategic nuclear weapons the United States and the USSR could deploy.[200] It also allowed for an intrusive monitoring regime to verify the treaty's compliance. For example, each side would be given open access to some previously encrypted signals. A new inspection regime was instituted, too. American and Russian weapons experts would be deployed on site at missile production facilities in each other's country to monitor the treaty's numerical limits on missile production, and each side could engage in surprise inspections at weapons storage and other locations to ensure extra missiles were not being stockpiled illegally. All told, these new provisions made nuclear weapons development, production and deployment more transparent. In a world of greater clarity and budgetary concern, it's possible that Rosman's services might have no longer been needed.

Presidents George H.W. Bush and Michail Gorbachev signed the START-I Agreement on July 31, 1991, in Moscow. *Courtesy of the National Archives.*

Regardless of the precipitating issues that led to Rosman's demise, Rosman Research Station probably was a victim of its own success. Between 1981 and 1985, the site was operating on a shoestring budget and at very little cost to the DoD. Then hundreds of millions of dollars in upgrades and increasing numbers of personnel were added to RSS, which led to a dramatic rise in yearly operations and maintenance costs. This made Rosman a highly visible line item in the DoD's budget—and because of it, a likely target for the BRAC program.

Rosman was owned by the NSA—that is, it was not located on a military base or on foreign soil, subject to the whims of other U.S. agencies or a host government. This made the closure decision easier, because the agency could act quickly without lengthy consultations with other stakeholders. In horse trading Rosman's future, the NSA could garner additional kudos for being a team player and perhaps use Rosman as a chit that allowed the agency to shield other potentially vulnerable facilities from the budget axe.

Locals believe the competition for survival played out behind closed doors in the U.S. Congress, where North Carolina was pitted against another state with a signals intelligence collection site about four hundred miles to the north: Sugar Grove, West Virginia. In making the rounds with Congress, NSA director McConnell would have been made intimately aware of the political headwinds and bent to the concerns of powerful politicians. In the competitive arena of base closures, North Carolina's delegation was no match for West Virginia's formidable senator Robert Byrd, whose five-decades-long career and position as a three-time chairman of the Senate Appropriations Committee gave him powerful leverage to steer federal programs and dollars to his home state.

ACT III.

ENTER THE PISGAH ASTRONOMICAL RESEARCH INSTITUTE (1998–PRESENT)

I saw that our country wasn't doing a good job with science education and decided to do something about it.
—Don Cline

The cities of Brevard and Rosman, indeed all of Transylvania County, was dealt a severe blow when, on March 31, 1995, the U.S. government, for the second time, closed its doors to the Rosman site. The DoD's Rosman Research Station was a major employer in Transylvania County, pouring about $5 million per year into the local economy. While contractors were given the option to relocate to another U.S. signals intelligence facility in Germany, those with strong family ties to Transylvania County, some going back many generations, stayed put. In doing so, they chose family over salary, as many former employees never attained the income they received as DoD employees.

Though NSA stopped its signals activities in March 1995, it didn't immediately turn the facility back to the U.S. Forest Service, as there were "many steps required to terminate cryptologic activity."[201] Rosman's unique DoD missions were reassigned to other cryptologic ground stations with similar access to foreign satellites and signals, and on-site personnel immediately began the hard work of deconstructing and disposing of equipment. Nineteen of the site's twenty-three major satellite dishes were disassembled bolt by bolt and hauled away, along with specialized signal collection and processing equipment, to NSA headquarters in Fort Meade,

Maryland, or to field sites in Virginia, California and England. The four satellite dishes the DoD left behind probably were altered to only receive, not transmit, signals and then mostly in lower satellite frequency ranges. The DoD removed other transmitting antennas and some that probably operated at much lower and higher frequencies. Other usable electronics were transferred to the Defense Logistics Agency and General Services Administration. The NSA donated personal computers—with the hard drives wiped clean—to area schools, and the site's firetruck was given to help establish the Balsam Grove Volunteer Fire Department. Disposing of fourteen years' worth of top secret files represented one of the organization's most daunting tasks. The shredder building was put to the test, generating almost four tons of paper shavings, which were delivered to Transylvania County for recycling.[202]

The DoD agreed to assume financial custody of the site for an additional eighteen months, until September 30, 1996, after which, if the site did not acquire a new tenant, it would revert to the Department of Agriculture's Forest Service. But neither organization wanted to contemplate that possibility. The Department of Agriculture's rules stipulated that the property given back to the forest service had to be returned to its original state—it had to be a forest. This meant that all buildings, pavement and concrete would have to be torn up, leveled and removed. This included the remaining infrastructure: water, oil, sewer and other hazardous waste in pipes, tanks and power generators; electrical wiring and optical cables—in short, everything the DoD inherited or installed in its massive facility upgrade would be uprooted and hauled away, and the site would then be planted with trees. The clean-up costs for the DoD would be huge, and the resources the forest service would need to oversee this activity, would be overwhelming. In preparing for the worst, the DoD, in January 1995, conducted an environmental assessment.

While neither the DoD nor the forest service wanted to undertake site reforestation, the challenge of finding a new owner within an eighteen-month window would prove daunting. The forest service had to market the facility to a select buyer with very narrow and specialized interests: someone or an organization interested in a rural, partially dismantled and aging ground station with thirty-year old antennas. Most importantly, the person or organization would need pockets deep enough to rebuild the site most likely for commercial or educational purposes. But beyond the plans and qualifications of the prospective owner, there were other wrinkles that made any transfer a nearly impossible proposition: the U.S. government's rules stipulated that the Rosman Research Station could not be purchased

One of PARI's many optical telescopes that are used to explore distant galaxies. *Courtesy of PARI.*

Ground was broken for NASA's most advanced ground station in February 1962. *Courtesy of the National Archives.*

Top: Orbiting solar observatory-class (OSO) satellites offered the first extended scientific study of the sun. The launch of OSO-8 took place on June 21, 1975. *Courtesy of NASA.*

Bottom: Senior Rosman engineer Joe Collins beside the ATS-6 satellite on display at PARI. *Courtesy of PARI.*

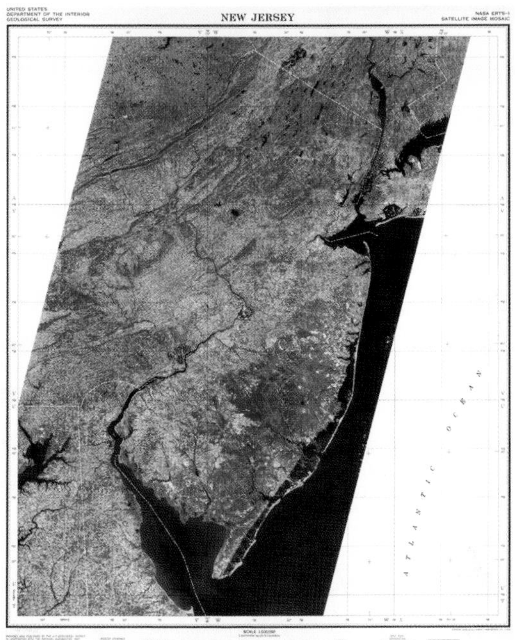

Above: This first full color image of the entire Earth was taken by the ATS-3 satellite, commanded from NASA's Rosman station, on November 10, 1967. It was the iconic image on the front cover of the fall 1968 edition of the *Whole Earth Catalog*. *Courtesy of NASA*.

Left: NASA's Rosman Station issued commands and collected signals from the first Earth resources technology satellite (ERTS). This image of New Jersey from ERTS-1 was taken on October 10, 1972. *Courtesy of the U.S. Geological Survey.*

Above: The Rosman Station supported the Skylab Orbital Workshop, which was launched 270 miles above the Earth on May 14, 1973. *Courtesy of NASA.*

Left: A Soviet stamp honoring the Molniya-1 satellite, which was launched in elliptical orbits around the Earth (1966). *Courtesy of the Post of the Soviet Union.*

Left: Students explore the tools of astronomy. *Courtesy of PARI.*

Below: Visitors learn about different spectra of light, split in a prism. *Courtesy of PARI.*

PARI's president Don Cline explains meteorites to PARI students. *Courtesy of PARI.*

PARI visitors observed a total eclipse in August 2017 using special solar viewing glasses. *Courtesy of Tom Tate/PARI.*

An aerial view of PARI today. *Courtesy of Don Cline.*

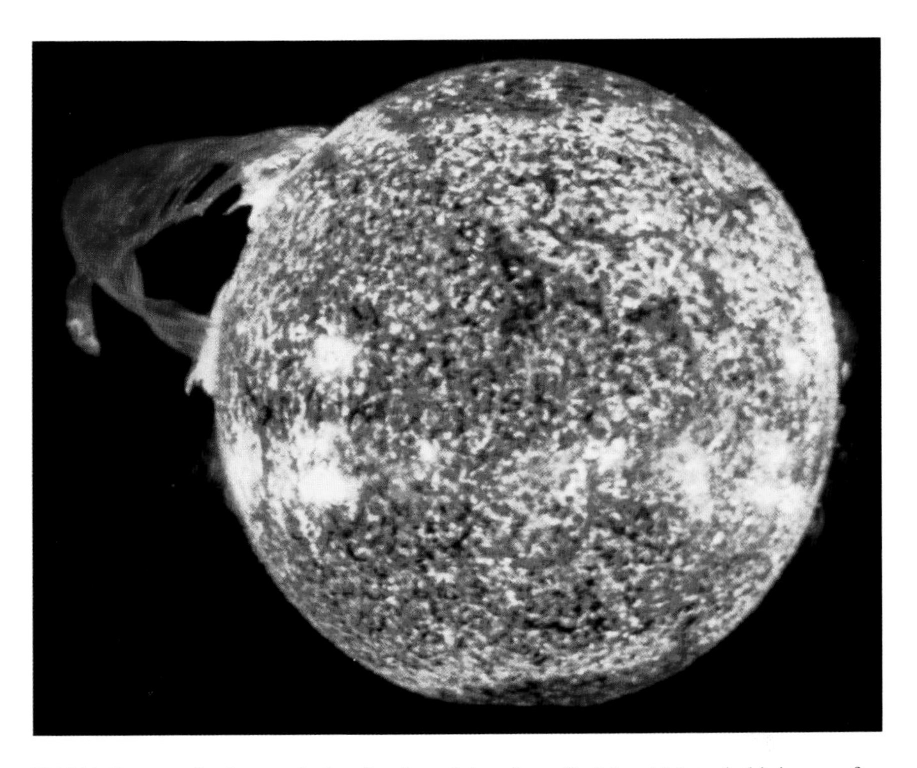

NASA's Rosman Station tracked and collected data from Skylab, which took this image of a solar flare. *Courtesy of NASA.*

Left: Rosman's second eighty-five-foot-wide antenna, constructed of steel in 1966, weighs over four hundred tons. *Courtesy of PARI.*

Below: PARI's museum hosts a collection of crystalline minerals that collect ultraviolet light and release the energy as brilliant luminescent colors. *Courtesy of PARI.*

Top: The launch of the ATS-3 satellite from Cape Canaveral, Florida, in November 1967. *Courtesy of NASA*.

Bottom: PARI is home to a variety of optical telescopes that allow visitors to view the nighttime sky. *Courtesy of PARI*.

Under the DoD/NSA, antenna radomes provided environmental protection and kept prying eyes from identifying the targets of intelligence collection. *Courtesy of PARI.*

PARI offers private tours and hands-on learning opportunities. *Courtesy of PARI.*

Left: Many evening programs give visitors a rare glimpse of meteors and other celestial bodies. *Courtesy of PARI.*

Below: PARI is nestled in the picturesque mountains of western North Carolina. *Courtesy of PARI.*

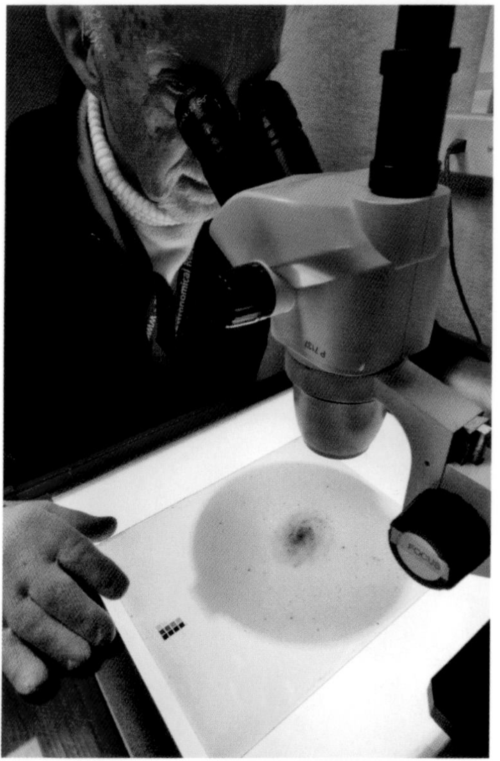

Top: Instructors at PARI teach STEM principles that have practical, real-world applications, like robotics. *Courtesy of PARI.*

Bottom: The institute's archive of historic glass plates offers researchers insight into the earliest days of our universe. *Courtesy of PARI.*

Top: Tunnels are the subject of PARI myths but were constructed by NASA to run power and control cables to satellite dishes. *Courtesy of PARI.*

Bottom: An early aerial view of NASA's Rosman Space Tracking and Data Acquisition (STADAN) facility. *Courtesy of Jo Ann Jackson.*

Top: Rosman had major responsibilities for the Orbiting Astronomical Observatory (OAO), the first telescope in space and the ancestor of the James E. Webb Space Telescope. (Image depicts the launch of OAO-2.) *Courtesy of NASA.*

Bottom: PARI uses optical and radio telescopes for students to study galaxies like the Milky Way. *Courtesy of PARI.*

Top: The first of Rosman's two eighty-five-foot-wide dishes was rushed into service to track and collect data from observatory-class satellites placed in polar and elliptical orbits. *Courtesy of the National Archives.*

Bottom: Rosman's largest dish, 85-I, could hear signals less than five watts (the power of a child's night light) two hundred thousand miles into space. *Courtesy of the National Archives.*

Left: An image of a galactic center taken from a PARI optical telescope. *Courtesy of PARI.*

Below: The PARI campus at night. *Courtesy of PARI.*

AVAILABLE

Satelite Communications & Data Center
±104,000 Square Feet on 202.35 Acres
For Sale

State Road 1326
Brevard, North Carolina
Pisgah National Forest
Transylvania County, North Carolina

For Information Contact:
Tom Evans
704-376-7979

Some antenna structures pictured above are no longer in place.

The sales brochure for the DoD's Rosman Research Station, which contained the frequency ranges for the four dish antennas that DoD left behind. *Courtesy of PARI.*

outright. A prospective owner had to find land or buildings the forest service wanted, purchase and then trade them for the now-202-acre Rosman site. Finally, the U.S. government had a say in the deal—it had to be blessed by both houses of Congress.

In June 1994, five months before NSA's last mission was shut down, the DoD, in conjunction with the forest service and army corps of engineers, hired consultants to help identify the market of potential buyers. The consultant's final report, issued in March 1995, came to the lofty conclusion that the Rosman Research Station "can and should be converted from a federal facility to a state or private facility that will benefit the citizens of western North Carolina, other parts of the state and the nation."[203]

Still, finding the right buyer would prove elusive, given the many hoops that had to be jumped through. Predictably, there were many false starts along the way. Early on, the NSA claimed, optimistically, that one potential customer, Texas Joint Ventures, wanted the site as an "uplink to the information superhighway."[204] This and other early prospects didn't bear fruit, and the site languished. After the DoD's September 30, 1996 custody commitment slipped by, the forest service assumed full responsibility for the site's maintenance. It immediately began implementing draconian cost-savings measures, turning off all but the most essential infrastructure and paring the staff to a minimum. In fact, only one person stayed on to maintain the two-hundred-acre facility. It was the man who had helped develop and clear the site for NASA and worked there for the space agency and through the DoD years: local resident Thad McCall.

FINDING THE RIGHT OWNER

After the DoD abandoned the site, the Pisgah National Forest's district manager and self-acknowledged landlord Art Rowe was under intense pressure to find a new tenant for the site:

> *It didn't make sense for the DoD to rip up all that concrete, so Thad McCall and I began to show the property. There were a lot of silly prospects. One woman wanted it because she liked the picnic area. Another potential buyer needed time to get the money. He said he had to sell a plane in Spain and gold certificates in Antigua.*
>
> *We didn't have a lot of time. After the DoD took off, we had to get rid of the property fast. The cost of the power bill alone was in the hundreds*

ALL IN THE FAMILY: THE McCALLS' SIX DECADES OF SERVICE

Four generations of the McCall family have worked at the Rosman site and three McCalls were the first hires. Fleming "Jahu" McCall and his then-eighteen-year-old son Thad were hired by NASA to cut an access road, now known as PARI Drive, that Thad's brother "Buster" then graded. Each would work for decades at the Rosman site.

Longtime employees Thad and Buster McCall. *Courtesy of PARI and the author's collection.*

"Jahu" continued to serve as a diesel mechanic and later as a Pinkerton guard through the DoD years; "Buster" in site construction, first digging the footers for the big antennas and later driving heavy equipment. Then more McCalls joined in. Thad and Buster's brother Ted was a carpenter at the site and as the years passed, Thad's son Brad assumed his father's responsibilities as head of security and facility maintenance. Buster's daughter Ann Daves serves as PARI's administrator. Finally, Brad's son Dylan works maintenance—the fourth generation of McCalls to call PARI home.

Though Buster McCall worked off and on at the site for sixty years, Thad McCall had the longest continuous service at the site: fifty-four years through NASA and the DoD—and later with PARI.

Don Cline was quoted as saying that he would not have assumed responsibility for PARI if Thad McCall wasn't "part of the deal." So valuable was Thad McCall's knowledge, that Don signed the papers acquiring the site only after Thad agreed, in a handshake deal, to continue working at PARI for at least a decade.

of thousands of dollars. We let go all of the maintenance folks except for Thad, and still, the costs were through the roof. The forest service didn't have the money. Of course, there were things that we had to keep running, like the sump pumps. Every time we had a potential buyer, Thad and I had to turn on the lights, and after they left, we had to go around and turn them off. It took forty-five minutes each time. The pressure was intense to pull the plug.

Nervousness over the lack of a viable prospect led the forest service to ask consultants in early 1997 to update the original marketing report, which would now take into account the equipment the DoD had removed in March 1995 and "excessed to other SIGINT facilities." The forest service also made clear that consultants needed to take a broader view of potential buyers. While the updated report continued to stress that a prospective buyer would likely be in telecommunications or education it also concluded the site's "one hundred thousand square feet of high-quality floor space...could support light manufacturing, office, sales, marketing or engineering [businesses]."[205]

The forest service would have to sit on the property, grit its teeth and pay the bills for nearly three years before a white knight arrived at Rosman's front gate with the resources and vision to make good use of the facility.

A STAR IS BORN: A NEW OWNER AND MISSION FOR THE ROSMAN RESEARCH STATION

J. Donald Cline always had a love of science. When the Statesville, North Carolina native was nine years old, he pulled an eyepiece from an old microscope, purchased a two-and-three-quarter inch objective lens and built his first telescope. His interest piqued, at age eleven, Cline ground a six-inch mirror for another homemade telescope. His love of science and technology continued to blossom throughout his successful career as a Bell Labs engineer and later, in 1977, with his ownership of Micro Computer Systems, a manufacturer of telephone testing equipment. He sold the company in 1995 and turned his attention to his earliest love, astronomy.

PARI's founder and president, Don Cline. *Courtesy of PARI.*

RADIO TELESCOPES PROVIDE UNIQUE INSIGHTS INTO THE LIFECYCLE OF STARS AND PLANETS

Optical telescopes collect waves from the visible light spectrum for observation but can only tell us part of the scientific story. Radio telescopes collect a much broader set of wavelengths, outside of what we can see with visible light, and provide rich detail about planets, galaxies and other celestial phenomena.* Because humans are visual creatures, seeing or imaging is an important part of astronomy, regardless of the kind of waves collected. Radio waves can be converted into image data.

Radio signals from stars, galaxies and a variety of space phenomena provide insight into the heat, chemistry and movement of celestial bodies millions of lightyears away.† Since the light waves now reaching the Earth were released billions of years ago, using a radio telescope is akin to peering through a time tunnel, revealing clues about the earliest origins of our solar system and the galaxies beyond. With the collection and assessment of these signals come immense possibilities for discovery, scientific research and the education of future generations of young scientists.

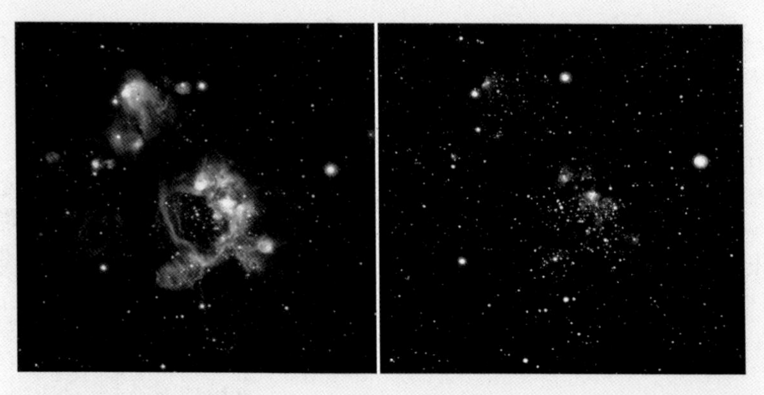

Optical telescopes use different lenses to capture more information from the visible light spectrum. Radio telescopes collect a different set of information from radio waves invisible to the human eye. *Optical images from the same satellite galaxy in the Milky Way, 157,000 lightyears from Earth, using different lenses; Las Campanas Observatory in Chile (December 2, 1978). Courtesy of PARI.*

---

* A third type of telescope emits radar signals that bounce off surfaces and can be used to map planetary terrain.

† A lightyear, the distance light can travel in a year, is 5.88 trillion miles.

Now, the engineer turned entrepreneur was searching for an antenna to bring back to Appalachian State University's Dark Sky Observatory. It already had a thirty-two-inch optical telescope, but Cline, a member of the university's Arts and Sciences Advancement Council, wanted a big antenna for its deep space astronomy program. University professor Dan Caton brought the forest service's real estate brochure to Cline's attention. "The government closed down the site," Cline recalled Caton telling him. "And they have some surplus antennas." Together, they made a bargain-hunting expedition to the Rosman Station in search of an antenna the government no longer wanted.

A dish antenna is a tool with many applications, and Cline had a creative idea: the government antenna, cutting edge for its era and earlier used to collect radio waves from scientific and foreign communications satellites, could be repurposed to push beyond the realm of man-made objects to collect electromagnetic waves emanating from the deepest corners of space.

DECADES-OLD EQUIPMENT RECEIVES A NEW LEASE ON LIFE

We were about to give up. Then Don Cline showed up. He was looking for an antenna to move to Appalachian State University, and he heard that we had one in Rosman. Well, when he saw this site, it blew him away. Thad took Don around while his wife, Jo, was outside, reading in the sun. After the tour, I said, "I think we have it."
—Art Rowe

When Cline and Caton visited the Rosman site, a daunting reality quickly set in. Dismantling and transporting the satellite dish they wanted would prove to be more challenging than they originally thought. Staring straight up at the 125-foot-tall, 85-foot-wide, 325-ton dish antenna, Don said, "I don't think we can move this one." But that got him thinking. A site with four dishes was available. Perhaps he could interest other institutions of higher learning into turning the entire facility into a hands-on educational enterprise.

"The United States wasn't doing a very good job with science education," Don would later say. "Few [young people] had a hands-on opportunity in science research." Cline had seen this problem up close as an entrepreneur and business owner, when he was unable to fill a number of science vacancies in his start-up business because new graduates lacked the practical experience needed to make the leap from classroom science theory to real-world applications. As part of his educational vision, Cline wanted students to experience the tactile thrill of discovery that astronomy could offer—to stimulate eager minds and offer a springboard to careers beyond astronomy in science, physics and mathematics. It was Cline's hope, too, that colleges and universities would use the facility and offer the continuing financial support that would make the institution self-sustaining within a short period of time.

But at this point, Cline's vision was just a dream. There were a series of hurdles to overcome, any one of which could upset the deal. Finding financial backing was just one of the steps; then came identifying the land of interest to the forest service, negotiating the swap and then, potentially, the most formidable stride in the long process, the approval of both houses of the U.S. Congress. Undaunted, Don's first stop was to the offices of the presidents of the sixteen universities that comprise the University of North Carolina system. "They were receptive [to participating]," Cline said, "but told me it would take five years to make it happen. But really, it would have taken ten or fifteen." Time was ticking away, and the forest service was on the verge of shuttering the site for good. As Don Cline saw it, he had only a few months to close the deal. He worked the phones, made personal trips and eventually collected several financial backers, but as the former Bell Labs engineer later recalled, "After I signed the papers, all of them disappeared."

Undeterred, Cline was willing to shoulder the burden of financing his expansive vision of an institution that, using optical and radio telescopes, would foster science education for future generations.[206] The next step was finding the land of interest to the Forest Service. Cline learned the government had several parcels it wanted, including 375 acres in the Pisgah National Forest with 7,000 feet of river frontage on the French Broad River adjacent to an established campsite but had no money to buy it.[207] Its ten-year option on the waterfront property was about to run out, and a developer, willing to pay four times the option price, was waiting in the wings. Cline acted quickly to purchase that parcel and two others for the forest service.[208] With the property that was up for the swap secured and the letter of intent signed, the sale entered a required ninety-day cooling-off period for the Environmental Protection Agency to survey the PARI site. It also gave area residents an opportunity to register their concerns about the prospective sale and time for politicians to look over the deal.[209] What remained seemed the most insurmountable of obstacles: the act of both

A happy day: Art Rowe turning the keys over to Don Cline. *Courtesy of Art Rowe.*

houses of Congress to approve the arrangement. But according to Cline, that turned out to be the easy part. The U.S. Forest Service, eager to ink the deal, handled the negotiations on Capitol Hill. With the support of the North Carolina delegation, Congress approved a line item inserted into an appropriations bill in early 1999.

> *It took nine months for the deal to go through. Giving Don the keys was one of the greatest days of my life.*
>
> —*Art Rowe*

FROM SURVEILLANCE TO SCIENCE: BUILDING BACK PARI'S EQUIPMENT AND INFRASTRUCTURE

Cline didn't sit back and wait for Congress to act on his proposal. After signing the letter of intent in June 1998, he began considering names for the former Rosman Research Station that would focus the site's mission, goals and aspirations. Don's friend and Raleigh, North Carolina marketing executive John Avant offered some possibilities: Pisgah Public

A star is born. *Author's Collection.*

Observatory, to highlight its goal of public accessibility, or Pisgah Forest Astronomy Institute, which would make the connection to education within a natural, "dark sky" setting. Eventually, over a dozen names were tendered for Don and the board for review, and in July 1998, one rose to the top of the list. Henceforth, the former Rosman Research Station would become the not-for-profit Pisgah Astronomical Research Institute (PARI). Its name underlined the importance of undergraduate and graduate research and public education.[210]

Acquiring the site and giving it a name were important first steps toward transforming Cline's vision into reality, but there were many miles to go before the deconstructed ground station now known as PARI could be turned into a functioning astronomical institution of higher learning. Cline quickly assembled a team of expert scientists to examine what he had acquired and determine how to best give his dream life. Avant recalled that "Don wanted it as a place where astronomers would want to come. But the physical plant at PARI was in shambles. It had been empty for nearly five years. Volunteers were rehabbing the facility, pulling cables and ripping up tiles, while work was continuing on restoring the four functioning radio telescopes. It was a mess."

VOLUNTEERS: THE LIFEBLOOD OF PARI

Volunteers have been an important part of PARI's story from its earliest inception. Whether they are involved in educational programs as night sky guides, assisting with public events or working on maintenance and landscaping, volunteers come to PARI for fun, camaraderie and personal enrichment.

My wife and I live in Florida and first took notice of PARI in 2002 through one of their publicized "Space Days." I've always loved astronomy and took an immediate interest. In those days, we'd have volunteer weekends two or three times a year, and between thirty and fifty folks would show up. In its infancy, the [site's] *needs were infinite. We were told, "Come on over and find something you'd like to do." It might be answering the telephone at the front desk, work on the Galaxy Walk or help with private tours.*

PARI relies on volunteers, like twenty-year veteran Joe Phillips, to fill a variety of roles. *Courtesy of Joe Phillips.*

PARI attracts volunteers from all walks of life. Some are fluent in science and can help operate or maintain the dishes, weather stations or seismometer. Others with trade skills help with plumbing or electrical work. Others are carpenters or plumbers. Some were literature majors and like working in the museum, researching the historical artifacts. Regardless of background, I think the common thread that draws volunteers to PARI is curiosity—about science, the stars and life.

—Joe Phillips

Upgrading one of PARI's eighty-five-foot-wide radio telescopes for a new mission: science education and research. *Courtesy of PARI.*

In surveying the site, the team found equipment of varying utility. Aside from the four functioning telescopes, the DoD spent hundreds of millions of dollars in upgrading the site's infrastructure, as its important national security mission couldn't tolerate the outages that accompany rural life. So, the DoD invested in a vast network of high-speed, high-capacity fiber optic cables and redundant "uninterruptible" power, water and sewage systems. But the high-security facility also had equipment of little value for a research and educational institution: four guard posts with ballistic glass, a three-lane pistol range, a compartmented floor plan with rooms separated by thick steel vault doors, a wide assortment of electronic and mechanical locks, disconnected red and green telephone lines used for classified discussions and a bunker-like conference room capable of producing white noise to prevent conversations from being overheard in adjacent rooms.

Sorting the useful from the useless equipment took some time, but according to Lamar Owen, PARI's chief technology officer, "When the DoD departed, they left behind a remarkable and mostly salvageable technical infrastructure." Eventually, Cline's engineers began bringing the site back online. Part of this revival required new motors for the largest telescopes. Under NASA and the DoD, some of the site's dish antennas tracked fast-moving satellites in low-Earth orbit. The celestial bodies under investigation by PARI required a set of finely tuned motors for precise coverage of objects millions of lightyears away that appeared to barely move at all. So PARI spent $3 million to adapt telescopes to slow-motion tracking. Cline also eventually acquired more than twenty optical telescopes of varying focal lengths and mirror sizes to add to the four existing and operational radio telescopes.[211]

THE DREAM IS REALIZED IN FITS AND STARTS

Restoring PARI took several years, but that gave the board time to think about its new mission. According to Mike Castelaz, PARI's former director of astronomy, science and education and early PARI hire, this meant addressing the overriding question: What do we want to do with PARI now that we have it?[212] Guided by its primary mission of undergraduate and graduate research and education, the PARI board surveyed its options. Board member Avant recalled, "One of the challenges is that it's difficult to become self-sustaining financially in education. We needed to have a regular funding stream."

About thirty undergraduate physics departments were within a few hours' drive of PARI, and Don Cline began looking for willing partners.[213] Soon after establishing PARI, he started working on existing relationships with the University of North Carolina (UNC) system, with plans to make greater use of PARI's telescopes. Cline knew UNC astronomy and physics professors were in a bind. The National Radio Astronomy Observatory's (NRAO) telescopes that were available for research were far from campus, had rigid usage requirements and were vastly oversubscribed. According to Wayne Christiansen, a then–astrophysics professor at UNC–Chapel Hill, "Time on a radio telescope is fought over. Young students are at the end of the bench and just don't get into the game."[214] But with PARI, the NRAO telescopes weren't the only game in town. UNC students would have opportunities for the kind of hands-on research and educational experience unavailable to their peers at other schools. PARI's unique capabilities would fill a special educational niche.

Working with the University of North Carolina system seemed a natural fit. In 2004, PARI board member Avant recalled, "UNC was big on the center concept," which offered an interdisciplinary approach to education. For PARI, this meant blending astronomy with mathematics and physics into its curricula, an arrangement that dovetailed nicely with its capabilities and mission. For the next two years, PARI worked with UNC on the plan, and it was on track, having successfully passed through the North Carolina General Assembly. It was a win–win. PARI would have a regular source of income that came with hourly dish usage, and UNC students would have unfettered access to PARI's telescopes for research and education. Under this arrangement PARI, associated with the UNC network, would host what would be called the Pisgah Astronomical Research Science and Education Center (PARSEC). All the details were falling into place—that is, until

"THE DAY THE RUSSIANS CAME TO PARI"

During Rosman's DoD's years, the site was off-limits to Soviet then Russian diplomats and military officials. But after PARI took over in 1998, the facility was opened up once again for domestic and foreign visitors.

> *One day in 2003, I got a call from a professor from Brevard College. Some Russians scientists were coming through town on a tour arranged by the Library of Congress, and he asked, "Would you mind if they came to visit PARI?" I said, "Sure, bring them in."*
>
> *There were about thirty Russians and two interpreters. We showed them around, gave them an overview of our STEM education program and played our PARI video. When we finished, I asked if they had any questions.*
>
> *One man stood up and began speaking in Russian. Now, I don't speak Russian, but it was clear that the longer he went on, the more agitated he became. After he stopped, the interpreter put her head down and shook it from side to side. In a low voice she said, "I'm sorry to say this, but this man claims what you're saying is all a big lie and that you are still spying on Russia from this location."*
>
> *"Well," I said, "there's nothing I can say to convince him otherwise, so I won't even try."*
>
> *—Dave Clavier, former PARI vice-president for administration*

the plan fell apart. In 2006, a new UNC president established a program where the university's sixteen institutions would compete for funding. The center concept, in which institutions worked cooperatively, each having separate responsibilities to complete central tasks, fell out of favor. Under the new UNC president, educational silos gained prominence at the expense of integrated learning. Established institutions had the inside track for educational training, and PARI was at a competitive disadvantage.

The demise of PARSEC as a source of funding was disappointing, but it led to a shift in emphasis, not mission. The board began, by necessity, to emphasize a more decentralized and broader approach to research and public education. It developed relationships with some forty leading institutions, including UNC–Chapel Hill, the University of Tennessee, the Georgia Institute of Technology, Duke and Clemson. And it asserted a more inclusive vision of education that incorporated Don's long-held view that science learning should engage all ages, "nine to ninety-three," or what Cline called "K to gray," with a special focus on preparing middle and high school students for further education and careers in science, technology, engineering and mathematics (STEM) fields. Specialized programs were offered for teachers, allowing them to earn credits required for recertification in science education as they brought new experiences in astronomy and other science disciplines back to the classroom.

A HANDS-ON APPROACH TO SCIENCE EDUCATION IN AND OUTSIDE OF THE CLASSROOM

According to Mike Castelaz, a key element of PARI's education program is to give students "a hands-on experience so they can see that science can be fun. If, afterward, they say, 'This isn't for me,' that's OK. It's all about having access to the world of science." Throughout the United States, there are few similar opportunities for grade school students to learn the principles of science using the instruments of astronomy. This represents the very core of PARI's self-defined role in education: to spark interest and to find and prepare the next generation of scientists for an increasingly competitive, technology-based world.

PARI reaches out to schools through age-appropriate educational experiences that meet the needs of schools and the state's educational guidelines. For grade school students, PARI's planetarium fills that niche. It's a large inflatable igloo that uses a star projector to place an image of the night sky on the interior dome. PARI classes complement school programs in science and, more broadly, with curricula in social studies, art and history. For example, one of more than a half dozen planetarium programs feature *The Stars of Lewis and Clark*, which tells the story of how the explorers of the Louisiana Purchase used the sun, moon and navigational stars to map their journey across the United States. Another called *Stars of My People* highlights the stories told in the stars from the perspective of Native peoples around the world.

Beginning in the 1990s, PARI's Bob Hayward traveled with the planetarium to schools throughout western North Carolina and northern South Carolina. After meeting with principals and teachers, he tailored his program to meet their needs and the state's science curriculum standards. Afterward, Bob and the teachers critiqued the event, homing in on its utility and increasing the value of the planetarium experience. In 2005, Hayward and Christi Whitworth, a former director of education at PARI, traveled with the planetarium and reached about ten thousand students. More recently, increasing financial pressure on public schools has changed the way the planetarium is used. According to Hayward, "School funding has tightened, so I'm not traveling as much." Currently, school groups can arrange a visit to PARI to explore the planetarium and PARI's varied educational programs.[215]

While the COVID pandemic placed a crimp in travel plans for the planetarium, PARI's management evolved its approach to give as many students as possible a hands-on experience with scientific equipment in their hometown school classrooms. In 2001, PARI became the first institution in the United States to give the public access and control of a radio telescope via the internet. Indeed, PARI's Smiley, the fifteen-foot-wide radio telescope antenna and separate forty-foot-wide antenna, have been operated this way

PARI's planetarium sparks students' interest in the nighttime sky. *Courtesy of PARI.*

EXPANDING SCIENCE EDUCATION TO INTERNATIONAL STUDENTS

PARI's capabilities in radio astronomy are capturing the attention of educators abroad. Students as far away as Australia have used the internet to control the Smiley radio antenna to support research projects. PARI is helping in other ways to educate international students in science fields.

One such effort is a planned cooperative program between PARI and the University of Puerto Rico. The island's huge one-thousand-foot-wide Arecibo radio telescope suffered a catastrophic failure when, on December 1, 2020, a main support cable broke, causing the entire antenna to collapse. It was a catastrophe for Dr. Mayra Lebron, a radio astronomer and professor at the University of Puerto Rico:

> *After the collapse of the Arecibo antenna, we began searching for new opportunities for a hands-on experience in radio astronomy. We heard about PARI through a colleague, who told us about its high-quality programs.*
>
> *We plan to establish a cooperative program that would offer our students opportunities using PARI's radio antennas, and we would support PARI, too, in data collection. Because our primary language is Spanish, we also could translate some PARI educational materials to reach a broader audience in schools and universities in Puerto Rico, the United States and beyond. It's a collaboration that would benefit us all.*

PARI offers a unique opportunity for students to conduct research by controlling radio telescopes using the internet. *Courtesy of PARI.*

to introduce students around the globe to radio astronomy. Giving middle school, high school and college students the ability to control a research-grade antenna is an empowering experience that many astronomy graduate programs cannot match. Among other projects, students have controlled Smiley to identify and study the composition of stars millions of lightyears away, map the rotation of our Milky Way and the movement of other more distant galaxies.

The stars are the gateway for PARI to help students explore science, not just the facts or the "what" to learn but how to learn—that is, students encounter the scientific process and develop problem-solving skills that can be carried beyond science to other disciplines.

For some middle and high school students, their learning continues into summer months. STEM astronomy and space exploration camps are grouped roughly by school grade and interest and are designed to inspire curiosity, passion and self-confidence. One of the more popular camps, *Mission Control: Martian Frontier*, blends science with other disciplines as students develop a plan to reach Mars. The camp calls on them to prioritize mission goals, choose a capable and diverse crew and navigate funding, ethical and other diverse challenges. It's a gateway for students to learn about astrophysics, engineering, rocketry, food science, biology and other disciplines as an integrated whole. It's taught by PARI staff and outside guests who have real-life experience working on these issues. PARI also created a more traditional learning experience in conjunction with Duke University and perfected it over the past two decades. Called *Above and Beyond*, the course focuses on discovery and immerses students in the world of space science and research, exploring the universe through the perspective of astronomy, physics and astrobiology.

> *One of my best memories was the last full day of summer camp. The campers were finishing up their research projects and presenting them. Then we gathered in the multimedia room and sat in a big circle and went over our memories of camp. The kids began to realize that camp was coming to an end, and tomorrow, they'd be going home. As we went around the circle, everyone had a favorite memory, and almost every camper started to tear up. They kept asking us if we were returning next summer. As a counselor, as cheesy as this sounds, I really felt like I was helping these kids. Knowing that I had a direct impact on their lives was a major reason I kept returning.*
>
> —*Matt Shelby, former PARI intern*[216]

PARI students and interns conduct a variety of self-directed science projects. *Courtesy of PARI.*

For middle and high school students looking for other traditional research experiences within an outdoor setting, PARI hosts one- and two-week-long space exploration camps taught in association with Clemson University. Among other activities, students build and launch rockets, learn how light is used to communicate with distant space missions, receive guest lectures and spend a day at Clemson, visiting university labs and meeting professors.

In an earlier offering called the Space Science Lab, students built their own antenna and receivers. Then, according to former education director, Christi Whitworth, students used their homemade antennas to explore research projects based on their personal interest. "We gave them the materials, tools and guidance and told the students, 'Now, go for it.' It was empowering. They developed and applied technical skills to a real-world, hands-on science project that excited them. And they learned to collect and interpret real data." This experience propelled a number of students to continue studies in astronomy and other STEM fields. According to Whitworth, each year, four or five students ask PARI instructors for letters of recommendation to gain entry into college and graduate school with a focus in science and astronomy programs.[217]

There is serious intent behind these practical experiences: students gain technical skills and knowledge and apply what they've learned to solve problems. Along the way, they develop the ability to think critically and learn the scientific research method while building teamwork and interpersonal and presentational skills.

• • •

PARI has a substantial impact on developing young scientists, training and helping them establish careers in STEM fields. The following is a reminiscence from Patricia Craig, a post-doctoral fellow at the Planetary Science Institute, whose time at PARI fueled an early passion for astronomy:

Former PARI student Patricia Craig now helps direct the Mars rover *Curiosity*. *Courtesy of Patricia Craig.*

> *As long as I can remember, I've been interested in space and space exploration. I grew up in a very rural part of central Florida. And by "rural," I mean* rural. *I lived in a two-story log cabin down a dead-end dirt road in a town of less than two hundred people. I could go out at night and see more stars than any of my friends who lived in town.*
>
> *It was at PARI, as part of a program sponsored by Duke, that, for first time, I genuinely got excited about science. I did my first scientific study at PARI, formulated a project, collaborated with fellow researchers, adjusted a project to accommodate unexpected glitches and kept crazy hours observing astronomical objects. My time at PARI gave me the tools I needed to build the rest of my career. Being at PARI was like being a "visiting researcher" at a university. I got a crash course in radio telescopes, used them in my research, enjoyed occasional outings and, at the end of two weeks, made a final presentation. Not only did I conduct my first real science project at PARI, but I forged lasting friendships.*

• • •

A COMMITMENT TO SCIENCE RESEARCH

Few realize that PARI holds a rare, century-old record of stars, planets and galaxies. Beginning in the 1880s, astronomers routinely recorded their visual observations of the universe on glass plates coated with an emulsion, a process advanced by Civil War photography. The plates, about the thickness of a windowpane, became the source for discovering and disseminating knowledge about celestial phenomena. It was an exacting and painstaking process that yielded a detailed record that, over time, matured our understanding about the lifecycle of distant stars, planets, galaxies and other bodies and their orbits, information that underpins astronomers' understanding of the universe that we rely on today. The glass plates represent a durable historical record of the night sky that, if stored properly, can last hundreds of years.[218]

Yet this valuable longitudinal register of the night sky is being lost. Over the intervening decades, plates began accumulating in observatories around the world and, gradually but inexorably, began taking up storage space. They became a chore to maintain, needing a low-humidity environment, occasional cleaning and, more recently, costly equipment and manpower to digitize the

PARI's glass slides are important historical records useful for astronomical research. *Courtesy of PARI.*

data to make it more useful. Cost, expertise and storage requirements often exceed what observatories can afford, which eventually leads to the destruction of many plates. Beginning in the 1990s, advances in digital photography and computing made recording images on glass obsolete.[219]

This irreplaceable snapshot in time that recorded the lifecycle of stars, galaxies and planets was being lost forever. "We're not time travelers," Mike Castelaz said, "so how do you go back in time to investigate the night sky except with the data you already have?" Don Cline, already well aware of the value of the plates and their continuing destruction, took the case to a 2006 meeting of the American Astronomical Society. He brought Castelaz and Wayne Osborn of Case Western Reserve University along to elevate the issue—not only to save the plates, but first, to make more astronomers aware of their existence and value in astronomical research. To dramatize their point, they made a presentation using a decades-old plate of the Andromeda Galaxy, the only record of what the galaxy looked like on that specific date. Then Osborn took out a hammer and brought it down on the glass, smashing it to bits. The crowd gasped, and a piece of irreplaceable scientific data was lost forever. Many astronomers in attendance agreed with Cline and his troupe that the plates were important records, but still, their observatories didn't have the space, funds or expertise to make good use of them.

To help solve this problem, PARI established an Astronomical Photographic Data Archive (APDA), a central repository for historic glass plates that observatories can't retain. The APDA has become a focal point for astronomical research, and with three digital scanners, PARI can make the data from these plates more widely available to researchers. New facial recognition software now in development can be modified to detect changes in stars and planets overtime, making this data even more valuable and accessible to researchers.

But glass slides offer more than a view of celestial bodies and their relative positions in the night sky at a fixed point in time. Beginning in the early 1900s, a second kind of glass plate was created when a prism was attached to the lens of an optical telescope and spectral images of the light from stars and reflected from planets were also recorded on glass. This spectral light tells us a lot about the stars and planets being captured. For example, a star's luminescence reveals what is going on inside its core, its temperature and chemical composition. Examining the same star over time can reveal different energy states that correspond to different stages in a star's lifecycle. Planets are not a source of light, but they reflect it, and viewing their luminescence through a prism tells us about a planet's atmosphere and

An image of the Andromeda Galaxy, 2.5 million lightyears from Earth, taken from a glass plate. *Image captured on November 25, 1943, at the Warner and Swasey Observatory, Case Western Reserve University. Courtesy of PARI.*

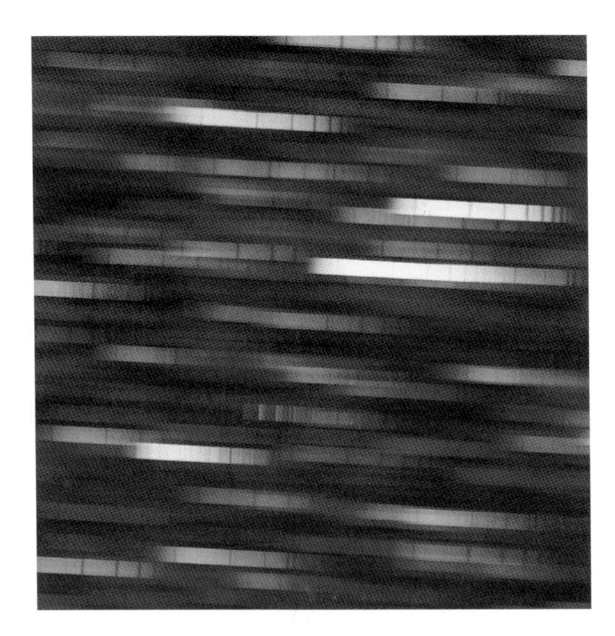

The luminescence of stars and planets seen through a prism. Spectral ribbons provide information on their compositions or atmospheres. *A glass plate image taken on October 19, 1971, at the Cerro Tololo Inter-American Observatory in Chile. Courtesy of PARI.*

whether it might be capable of sustaining life. To help catalog the stars and planets in the sky, PARI offers a citizen science project, giving the general public an opportunity to help classify stars and planets according to their spectral category.[220]

PARI's APDA is the second-largest plate repository in the world. The archive holds over 460,000 glass plates from 83 observatories around the globe, including historic collections from Yale and the U.S. Naval Observatory. Only Harvard University has more, with about 550,000. And while Harvard is no longer accepting accessions, PARI's doors are open for the nearly 2 million plates that remain at observatories scattered around the world.[221] In a few years, PARI will likely surpass Harvard's collection, but in the meantime, Don Cline boasts that the current collection is so extensive that you can "pick a point in the sky, and PARI has a plate of it."

My third summer at PARI, I tried a research topic I knew almost nothing about. I saw a news article about coronal mass ejections (CMEs) from the sun, which led to geomagnetic storms here on Earth. The worst CME in recorded history was the Carrington event of 1859. If this storm were to happen today and we weren't prepared, it could take us about fifty years to fully recover. Our electrical grids would shut down, astronauts in orbit would die, satellites would fail and much more. It would lead to trillions of dollars in damage.

I began to dig more into this topic and found the historical glass plates at PARI were perfect for this research. I'm looking back as far as 1903 to study images of the sun. Sunspots give rise to these CMEs, which then lead to magnetic storms that cause our electrical systems to fail. We found a correlation between sunspot and CMEs, and we're preparing a research paper for publication.

—*Matt Shelby*

MEETING THE PUBLIC'S VARIED INTERESTS AND NEEDS

A visitor to PARI might think that all there is to see are big telescopes or a beautiful starry night. While these are impressive sights, PARI also has a museum filled with NASA equipment and natural artifacts that also have their own stories to tell about PARI's unique relationship with its past and the natural world, whose secrets it still seeks to uncover. It's all part of what board member Zac Engle calls PARI's effort to "reach people where they are....There are things here to interest everyone." Tucked back in the main building, there's a hands-on exhibit of NASA hardware, including a 1960s vintage rocket engine from the Redstone program; an authentic *ATS-6* satellite, a sister of the *ATS-6* that Rosman guided in the 1970s; a one-third scale model of the Apollo lunar lander; and genuine equipment from the space shuttle *Discovery*. Farther back still is a world-class collection of meteorites from all over the globe, including one whose fall was recorded in 1492. Within the meteorite exhibit, there is a dark fluorescent tunnel where crystalline minerals collect light energy from ultraviolet wavelengths and then release the energy in the form of brilliant luminescent colors. NASA equipment, meteorites that tell the story of the origin of our galaxy, crystals that collect and reflect light of different wavelengths and the rest of the museum's content illuminates PARI's own story and fits easily with its mission of public outreach and education.[222]

During the NASA days, the Rosman facility operated regular tours and was open eight hours a day, six days a week. The 1960s were a simpler time, when space was a new frontier and just being on the grounds of a NASA site, looking up at large dish antennas, was enough to set imaginations rocketing into space. Today, the public's imagination is stirred by the images taken from the Hubble and Webb Space Telescopes before coming to PARI, and expectations are already high. They want to be moved even more deeply.

PARI IN THE PATH OF A TOTAL SOLAR ECLIPSE, 2017

Hundreds of visitors to PARI on August 21, 2017, had a ringside seat to watch what became popularly known as the Great American Eclipse. While it took three hours for the moon to cross the face of the sun, the total eclipse itself lasted little more than a minute. Residents and visitors from across the United States donned protective glasses to shield their eyes from the harsh rays of the sun and had an unobstructed view of the spectacle from within PARI's fence.

But beyond the entertainment value, there was serious science going on at PARI. According to Mike Castelaz, this was the first time that a radio telescope was in the path of a total solar eclipse. PARI trained its telescopes to look at the outer rim of the sun, called the corona, to detect hydrogen atoms, something that had never been done before. Different groups from NASA were also on hand to observe and take atmospheric measurements.

The path of the August 2017 solar eclipse. *Courtesy of NASA.*

A total solar eclipse. *Courtesy of NASA.*

While there will be other partial solar eclipses visible in North Carolina in the not-too-distant future, including one that will cross the United States from Texas to Maine on April 8, 2024, for many, the 2017 total solar eclipse was a once-in-a-lifetime event. North Carolina will next experience totality on May 11, 2078.

Unlike NASA, PARI as a nonprofit institution has a limited full-time staff, and since optical telescopes have daylight restrictions, PARI uses its resources in ways that optimize the site's capabilities to satisfy the public's interest. In 2019, the year before the pandemic, PARI had record attendance, with nearly eleven thousand visitors to the site, and touched the lives of an additional eight thousand with its off-campus programs. Social distancing and other restrictions associated with the COVID pandemic led PARI to downsize its programs and curtail its open gate policy, which reduced the number of visitors to PARI by half. With the easing of restrictions, attendance is back up, fueled by PARI's popular private daytime tours and evening events, when the telescopes are available for nighttime viewing.

> *One of the most underrated aspects of PARI is its night sky. It tends to get overlooked. During public and private observing nights that I would help run, there was never a night where people who came to see the stars weren't amazed. I loved seeing people from all ages and backgrounds—kids as young as five and adults as old as ninety—were fascinated by the same thing, the night sky and all of its mysteries. While I was at PARI, I was able to bring the sky a little closer to them and witness their pure excitement.*
>
> —*Matt Shelby*

PARI hosts many events during the year, including a well-attended annual "Space Day," and open house in the spring, that is free to the public and many regularly scheduled evening events tied to extraordinary celestial activities. For example, the Leonid Meteor Shower Camp-Out is a novel way PARI meets the public's varied age, knowledge and interest levels. For a nominal fee, visitors of all ages are invited to tent under the stars or sleep in a comfortable cabin room as the meteors light the evening sky. The public comes away from the event with a better understanding of the capabilities of PARI's optical and radio telescopes, because they experience the meteors, stars, planets, nebula and other celestial bodies on display in a pristine dark sky. According to PARI's director of software Tim DeLisle, "People are hungry for a hands-on experience," and they get it by taking control of the telescopes to explore the nighttime sky. These regularly scheduled programs fit neatly with the public service and educational mission of PARI. "The idea of 'build it and they will come' doesn't work," DeLisle said. "We discover the public's interest and build programs to meet them where they are."[223]

FUTURE CHALLENGES: SUSTAINING THE VISION OF SCIENCE EDUCATION

Despite the utility and availability of PARI's telescopes for academic research, sometimes it's difficult to maintain partnerships with other institutions. Part of this is due to the culture of astronomers, and some is due to the soft, uneven nature of grant funding. Finally, there's the personality-dependent nature of institutional associations.

Funding is a core issue, and sometimes, when institutions or individual astronomers want to collaborate, it's one-sided. They ask to use PARI's facilities and equipment without compensation and even want PARI to finance their research. This means that establishing collaborative relationships frequently rests on pursuing state and federal grants, which typically last for one to three years, so when funding dries up, so does the research and often the institutional relationship. The longest-lived collaborative associations are generally made with well-funded universities, but even this can be disrupted; they can be personality-dependent and falter when a champion of an established university collaborative program retires or moves to another institution. Successors don't necessarily have the same commitment to maintaining previous relationships. PARI has been successful in finding partners, but it requires constant work to grow the new relationships needed to preserve core educational programs. While funding from grants helps make PARI's programs self-sustaining, these funds often can't be used to operate, maintain or upgrade PARI's facilities.

> *What I really love is that we were able to continue to serve our regional students during the pandemic. The PARI staff put together take-home STEM kits, which included a thumb drive with video instruction and interviews with scientists so that our students could continue their education, even in areas where internet service does not reach their homes. The summer kits were such a hit that we offered them to the public schools during the next school year when students attended in person.*
>
> —*Randi Neff, Smokey Mountain STEM Collaborative*

PARI board member Zac Engle, an executive with substantial experience in nonprofit development, likens administering PARI to being a historical steward of an English manor. PARI is integral to western North Carolina, injecting more than $1 million a year into the local economy, and it offers great educational value to the residents of Transylvania County and the scientific community, but the facility also requires maintenance. In PARI's case, the DoD's legacy to PARI was its redundant and oversized infrastructure, which exceeds what the institution currently needs. For example, the DoD demanded two of everything, including two water tanks and pumps to ensure an uninterruptable supply of water. That means PARI now has multiple maintenance and equipment replacement requirements. This, according to Engle, is a challenge but also an opportunity. It requires upkeep but also offers a highly reliable infrastructure and growth opportunities for new partners.[224]

Even though PARI is a not-for-profit educational institution, it receives no federal funds for its educational programs. To cover costs, PARI management is aggressively pursuing commercial relationships with the public and private sectors. In doing so, it seems PARI is coming full circle. It's leveraging its greatest advantages, geography and geology, that were prized by NASA sixty years ago, to propel it into the future. Nestled within five hundred thousand acres of the Pisgah National Forest, PARI is certified as an international Dark Skies Park, one of only about 120 scattered around the world.[225] PARI remains a premier location for effective signals access to satellites placed in a variety of orbits, just the kinds of capabilities needed for effective ground-to-satellite tracking and command and control communications.

There are renewed possibilities, too, for rejoining NASA, this time as a commercial partner. The space agency has big plans through its Artemis program, beginning in the mid-2020s, to establish a long-term presence on the moon. PARI's large dishes could potentially serve as they had in the NASA days, when the station supported the Apollo program, by supporting spacecraft orbiting the moon.

Other fresh opportunities will continue to spur interest in PARI's capabilities. New satellite companies are springing up, and some are looking to establish a global satellite control and tracking consortium, in which a network of ground stations and personnel are "rented" as a package to space launch companies that need support but lack the funds to create their own collection of ground stations. Other companies are exploring mission-by-mission relationships with PARI. PARI's secure and internationally recognized "dark sky" location and robust power, water and server capabilities are an advantage when considering the facility as a data center, where the customer can capture and store or forward data using its own communications or PARI's robust fiber optic cable network. Likewise, PARI's campus, hardy infrastructure and access to the region's scientific and technical expertise could be transformed into a high-tech business incubator or conference center. While the opportunities for generating income seem limitless, PARI management's focus is crystal clear: business opportunities are not ends in themselves, they are a way to expand and grow PARI's mission of STEM education and research.

EPILOGUE

LOOKING BACK, MOVING FORWARD

Ultimately, the history of PARI is less about the institutions that called the Rosman site home than it is about the story of the people who worked there. From NASA to the NSA and, finally, PARI, the past six decades have witnessed an amazing ebb and flow of talented men and women brought together for a common mission and then dispersed, brought together again and then dispersed. A human tide not directed by the moon but by the demands of the times. Each new swell of employees brought remarkable contributions—some we see today in our daily lives, like cellphones and the internet. Some important contributions remain unseen, like living in a peaceful, secure and free country. And finally, there are contributions that we may not see for decades to come, like the new breakthroughs and discoveries achieved through science education.

All progress is rooted in experimentation, and today, PARI remains an experiment for which there is no model or guide. Still, the challenges it faces today are not that different from those of its predecessor organizations: maintaining its relevance and place in a rapidly changing world. NASA departed Rosman because satellites in space could now do what ground stations had done in the past. The DoD abandoned the site because, having won the Cold War, Americans were looking for a "peace dividend" and a reduction in the military's budget. Evolving technology and developing a sustainable stream of funding posed challenges then like they do today.

PARI's remarkable evolution from a twice-owned government facility to a not-for-profit institution is unique in this country and any other. This

The historical marker on Route 215 highlighting PARI's past. *Courtesy of PARI.*

transformation was possible because the tools of science were adaptable and rooted in the immutable laws of physics. They were molded and shaped to fit the evolving needs of each era, and there is no reason to believe they can't meet the new, more challenging era to come.

PARI has stayed relevant because of Don Cline, a man of vision and courage whose commitment to PARI and STEM education over the past two decades has remained steadfast. Transformation calls for leaders to not only accept change but recognize the opportunities it presents. Cline and his team are evolving PARI programs and capabilities to meet the public's fluctuating interests, the needs of students and shifting curricula while also offering greater opportunities for research. Developing new partnerships with NASA and private sector corporations in the burgeoning field of space commerce will help ensure that PARI continues to offer educational opportunities that change lives and spur scientific advances for generations to come.

The progress that comes from discovery is often revealed slowly and with patience, one scientist at a time. For students who are fortunate enough to call PARI home, these lessons are learned now as they have been for the past sixty years: through understanding the deepest secrets in the sky.

NOTES

Act I. Spacemen: The National Aeronautics and Space Administration Arrives in Rosman (1963–81)

1. George Reedy, aide to then-senator Lyndon Johnson, on hearing of the Soviet's *Sputnik-I* launch, in Robert A. Divine, *The Johnson Years: Vietnam, The Environment, and Science* (Lawrence: University Press of Kansas, 1987), 219.
2. This author uses the term *Soviet* ("citizen of the Soviet Union") when writing about the years leading up to the disintegration of the Soviet empire in December 1991. After that date, the country became known as Russia and its people, Russians; the largest ethnic group of the country.
3. The Soviet news agency TASS announced *Sputnik-I*'s transmission frequency (20–40 MHz [megahertz]) so amateur radio operators around the world could dial in to hear the satellite's telltale *beep*. Laika, "Barker" in Russian, survived launch but perished after the capsule overheated on its third orbit. Though *Sputnik*, in Russian, means "fellow-traveler," after its successful launch it now also means "satellite."
4. In another blow to America's prestige, America's first attempted satellite launch using the navy's Vanguard rocket ended in a televised fireball one month after *Sputnik-II* was launched. The country's first successful satellite, *Explorer-I*, was launched in January 1958. Between December 1957 and September 1959, eight of eleven Vanguard launches failed, leading then–Soviet premier Nikita Khruschev to dub Vanguard "the grapefruit satellite," possibly confusing it with another citrus fruit, the lemon.

5. The military's minimum tracking satellite ("Minitrack") network was developed to track Vanguard, a series of unmanned, Earth-orbiting scientific satellites. Minitrack became operational just three days prior to the *Sputnik-I* launch. A week later, station antennas were modified to track *Sputnik*, but the damage was done. Putting the best spin on the incident, the Eisenhower administration called the *Sputnik-I* launch a good "shakedown" exercise for Minitrack. Another embarrassed official called it "the wettest of dry runs" (Corliss, *Evolution*, 30). In honor of the Soviet achievement, it was said the Americans concocted a new drink called the "Sputnik": one-third vodka and two-thirds sour grapes. Constance McLoughlin Green and Milton Lomask, *Vanguard: A History* (Washington, D.C.: NASA, 1970), 188.
6. This fear was premature. While the Soviets first detonated a hydrogen bomb in August 1953, its first operational nuclear ballistic missile that could reach the United States, the SS-6 (*Sapwood*) missile, was not operational until 1960.
7. In the early 1960s, the U.S. manned space program was also playing catch up in the space race. Soviet astronaut Yuriy Gagarin was the first astronaut to fly a suborbital mission (April 1961), followed by the U.S. astronaut Alan Shepard one month later. Soviet cosmonaut German Titov first flew the first orbital flight (July 1961), followed seven months later by the American John Glenn (February 1962).
8. Even by the early 1970s, six of seven NASA launch vehicles were derived from military missile boosters (Levine, *Managing NASA*, 226).
9. The highly secretive National Reconnaissance Office (NRO), which was charged with developing, testing and producing reconnaissance satellites, was a joint CIA/DoD office. Its existence was declassified in 1992.
10. NASA first administered the scientific satellite and manned space flight missions (MSF) as a single program. In 1965, with the Apollo manned lunar landing program in full swing, the satellite and MSF programs were separated, and NASA established a third "deep space" mission. In 1972, three years after the historic Apollo lunar landing, NASA's budget contracted, and all three missions were combined and administered as one Spaceflight Tracking Data Network (STDN). Broadly speaking, the term *manned* was literally inaccurate, as Soviet Valentina Tereshkova became the first woman cosmonaut in June 1963. It was accurate for the United States until 1983, when Sally Ride became the first American woman in space.
11. This requirement is enshrined in the 1958 Space Act that established NASA (NASA, "History," https://history.nasa.gov/spaceact.html).
12. Kennedy's speech to a special joint session of Congress took place on May 25, 1961. NASA engineers said it would have cost twice as much to get to the moon if President Kennedy had not provided a clear deadline (Corliss, *Evolution*, 3).

13. Goddard Space Flight Center (SFC) in Greenbelt, Maryland, directed NASA's scientific satellite program. For the quote, see Wallace, *Dreams, Hopes and Realities*, 30.
14. For example, the first satellite Rosman would track, the Interplanetary Monitoring Platform (IMP), had an elliptical orbit with a perigee (the point closest to the earth) of 120 miles and an apogee (farthest point from the earth) of 121,000 miles. Satellites in geosynchronous orbits have an altitude of about 22,250 miles and travel at a speed that matches the Earth's rotation. If the satellite is placed at an angle to the equator, it passes the same spot above the Earth at the same time each day. A geostationary satellite employs a geosynchronous orbit in that it also travels at a speed that matches the Earth's rotation, but it stays continuously fixed at a designated position above the equator. This paper will use *geosynchronous* when referring generically to both sets of satellites. New satellite orbits required new command and tracking equipment and ground stations. For example, NASA established a site in Fairbanks, Alaska, to collect data from scientific satellites traveling in polar orbits (Corliss, *Histories*, 42–45; NASA, "Goddard's Rosman Site to be Dedicated Saturday," *Goddard News* [Greenbelt, MD], October 21, 1963, 1).
15. Minitrack used multiple ground stations to "hear" a satellite's transmitting beacon and then locate it. It's a process similar to human hearing, in which location is detected by the sound hitting two ears at different times. Vanguard collected data on the size and shape of the Earth and its air temperature and density.
16. The problem of tracking a satellite is inherently difficult. For example, if a pilot dropped a golf ball from a plane traveling at the speed of sound (761 miles per hour) at sixty thousand feet, the size and speed of this golf ball would be closely approximate to the size and speed of a satellite three feet in diameter at a height of three hundred miles. The acquisition problem is to locate the object and the tracking problem is to measure its angle, position and angular rate of travel with sufficient accuracy to alert other tracking stations and offer the exact position of the object and its expected time of arrival. New eccentric and elliptical orbits added layers of complexity to the acquisition and tracking problem (John T. Mengel, "Tracking the Earth Satellite, and Data Transmission by Radio," *Proceedings of the IRE* 44, no. 6 (June 1956): 755; Green and Lomask, *Vanguard*, 147).
17. For example, the Orbiting Astronomical Observatory-2, a forerunner of the Webb Telescope, returned data for more than nine years on the birth, death and life cycle of stars (NASA, "Space Science Coordinated Archive," https://nssdc.gsfc.nasa.gov/nmc/spacecraft/display.action?id=1972-065A).

18. "NASA Observes Eighth Anniversary of Rosman Tracking Station," *Transylvania Times*, October 28, 1971, 8; "Rosman Tracking Station One Year Old Monday," *Transylvania Times*, October 29, 1964, 4.
19. The $1 billion was dedicated for ground stations across all NASA space flight missions (Corliss, *Histories*, 3).
20. Corliss, *Histories*, 63; "Eighth Anniversary," *Transylvania Times*, 8.
21. NASA began using the acronym STADAN in 1964 to describe the new network of ground stations with upgraded equipment. In 1963, only three of STADAN ground stations were equipped with the largest, eighty-foot-wide dish antennas: Rosman, Fairbanks and Orroral Valley, Australia.
22. NASA had selected the Fairbanks site four months prior to Rosman. Eventually, other isolated sites with favorable geography, including the Mojave Desert (California), were added to the STADAN network.
23. Later, when NASA began using an eighty-foot-wide dish for deep space experiments, loudspeakers would blare across the site, asking employees to shut off their car motors, and travel within the station was halted (Janice C. Roller, "Rosman: A NASA Satellite Tracking Station, 1963–1981," unpublished photocopy, n.d., 8).
24. While the area is popularly known as the Tennessee Bald, its historically correct name is Tanasee Bald, named for a mountain near the Blue Ridge Parkway in western North Carolina. Webb previously served as the director of the Bureau of the Budget from 1946 to 1950 and the undersecretary of state from 1950 to 1952. These two positions involved considerable contact with national security agencies and required intimate knowledge of classified information and programs (see act II for more on NASA/Rosman's relationship with classified DoD programs).
25. Proximity to Goddard Space Flight Center in Greenbelt, Maryland, was critical due to the requirement of shipping magnetic tape for rapid processing and analysis. See "Rosman, North Carolina: Satellite Tracking and Data Acquisition Facility," in *Facility Brochure* (Washington, D.C.: NASA, n.d.), 2.
26. Jackie Langille, geologist, UNC-Asheville, email to the author, "A Question About PARI's Geology," May 4, 2021.
27. "Upper County Linked Electronically with Canberra, Australia and Alaska," *Transylvania Times*, October 29, 1964, 1.
28. Art Rowe, former Pisgah National Forest district manager, interview with author, Pisgah Forest, NC, April 6, 2022.
29. Contrary to its name, Fort Valley was never the site of a U.S. government fort.
30. Edmund C. Buckley, director of tracking and data acquisition, to NASA associate administrator, "Fiscal Year 1962, Project 3379, Rosman Data Acquisition Facility," (unpublished), February 14, 1962.

31. NASA, "Key NASA Tracking Site," 1.
32. Ibid., 1; December announcement: Congress, U.S. House, "Tracking Site, 1 Announcement: House Committee on Science and Aeronautics, Aeronautical and Astronomical and Aeronautical Events of 1961, 87th Cong., 2nd sess., June 7, 1962, 74.
33. Of the original fourteen STADAN stations in 1964, only three were modern data collection sites: Rosman; Fairbanks, Alaska; and Orroral Valley, South Australia. The other eleven were Minitrack stations. Construction on the Fairbanks site's eighty-five-foot-wide dish was completed a few months prior to Rosman's. Quote from Associated Press, July 1, 1963, in Jon Elliston, "Big Brother's Legacy: Spy Outpost Comes in From the Cold," *Mountain Xpress*, June 2004, 6.
34. *Facility Brochure*, 2.
35. Roller, *Tracking Station*, 6.
36. Some abandoned homesteads dotted the six-hundred-acre site, including one from the nineteenth century known locally as the "Old Becky Place," after its former owner Becky McCall. No one claimed ownership of the whiskey stills found on site, but if asked, locals called them "government stills" (Janice Roller, "Rosman: A NASA Satellite Tracking Station 1963–1981," unpublished manuscript, n.d.).
37. Tsiao, *Read You*, 36; "Big dish…energy," quote from: "Manager Named—Satellite Tracking Station Will Be Pivotal Facility," *Transylvania Times*, July 3, 1963, 1, 3.
38. Edmund C. Buckley, director of tracking and data acquisition, to NASA associate administrator, "Communication Satellite Ground Support Requirements Relative to Rosman II Station Plans," LS, January 4, 1963.
39. Properly outfitted eighty-five-foot-wide dishes could be tuned to either track satellites or receive their signals. The exacting requirements of tracking manned flights required large dishes. In scientific satellite programs eighty-five-foot-wide dishes were used for data collection. "Site to be Dedicated," *Goddard News*, 2.
40. NASA required a network of ground stations, because each station could only track or collect data when the satellite was in its field of view. After a satellite passed over the horizon, it was handed off to the next ground station. The number of STADAN sites fluctuated over time; new sites were added, and others were removed based on advancing satellite technology and orbits, budgetary constraints and the politics of the host country (Corliss, *Histories*, 60).
41. "NASA Tracking Station in Tar Heel Mountains," *Chowan Herald* (Edenton, NC), May 3, 1979, B1.
42. Tsiao, *Read You*, 39. The GRARR had precise capabilities. For example, it tracked the orbiting geophysical observatory-class satellites (OGOs) to an

accuracy of sixteen yards in distance and four inches per second in speed at altitudes up to 650 miles ("Another First in Space Is Noted at Tracking Station," *Transylvania Times*, September 17, 1964, 1).

43. To bolster satellite command and control, Rosman received six advanced satellite automatic tracking antennas (given the unfortunate acronym SATAN), three for receiving satellite data and three for issuing commands—two more than any other STADAN ground station. Rosman's SATAN antennas were used by Goddard Space Flight Center in what was believed to be the most distant "fix" ever accomplished when, in January 1967, command signals were sent to an "electric screwdriver" to restore power to *Explorer-33* in orbit 292,000 miles from Earth (over 50,000 miles beyond the moon). This maneuver saved the spacecraft from a power blackout and almost certain mission failure ("Satellite Power Fixed From 292,900 Miles," *Goddard News Roundup*, February 3, 1967, 8). NASA STADAN ground stations typically had between one and four new SATAN antennas. In 1966, Rosman had six, though two would be removed in 1969. Rosman used a three-tier approach to forward data to Goddard SFC in Greenbelt, Maryland. Information was reviewed on site, and the most important information was sent via a microwave link to Goddard SFC directly; the next priority data was flown via Asheville's airport to Maryland; the routine data was stored on magnetic tape and trucked to Goddard. To lower costs and increase efficiency, beginning in 1966, ground station computers used new data compression techniques to reduce redundancy and improve satellite data quality. By 1969, the STADAN was collecting two and a half times the volume at a much higher quality than the data obtained in the early 1960s (Corliss, *Histories*, 63).

44. "Satellite Tracking Station Work Schedule Announced," *Transylvania Times*, July 5, 1962, 1. Full operational capability need not coincide with a facility's completion or a site's public dedication. NASA expected Rosman to be fully operational in May 1963, barely a year after ground was broken, but the dedication ceremony was postponed to October due to delays in installation of electrical equipment. See Edmund C. Buckley, director of tracking and data acquisition, to NASA associate administrator, "FY1962 Construction Estimates; Rosman Data Facility Project 3379," LS, February 14, 1962; "Manager Named," *Transylvania Times*, July 4, 1963, 1, 3.

45. "By 1962, trailers were installed in Rosman, Tananarive and Carnarvon, Australia to support numerous Goddard satellite programs—to provide range-rate data, whose accuracy would not be surpassed until the use of lasers a decade later." The first observatory-class satellite, the Orbiting Solar Observatory-1 was launched on March 7, 1962. For information on the van's arrival, see Tsiao, *Read You*, 39.

46. The IMP, associated with the Apollo program, was the first satellite to use integrated circuits. It investigated cosmic rays and solar winds outside the Earth's magnetic field. The IMP was placed in a highly elliptical orbit that would give Rosman's dishes a good workout; it had an apogee of 121,000 miles and a perigee of 120 miles ("Rosman NASA Station Head Describes 'IMP' Satellite," *Transylvania Times*, December 5, 1963, 1).
47. "Manager Named," *Transylvania Times*, 1, 3; information on M. Gary Dennis is from Bill R., former NASA and DoD employee, interview by author, Balsam Grove, NC, October 30, 2021.
48. Rosman's first eighty-five-foot-wide antenna was operational in July 1962, about four months after ground was broken at the site (Coreless, *Histories*, 318).
49. NACA, the National Advisory Committee on Aeronautics, became NASA in 1958.
50. "Rosman Dedication Ceremony Precedes Site's Open House," *Goddard News* 5, no. 11 (November 4, 1963): 1.
51. Transylvania County Board of Commissioners, "Welcome Visitors to Transylvania County," *Transylvania Times*, July 7, 1966, 16.
52. NASA, "Key NASA Tracking Site," 2.
53. The data collected by these satellites was consequential. Severe space weather, for example, could expose astronauts to high levels of radiation, damage electronic equipment and interfere with space and terrestrial communications.
54. The television infrared observation system (TIROS) was an early weather satellite. Nimbus, an experimental weather satellite, became the platform for ERTS, the Earth Resources Technology Satellite, later renamed Landsat, which hosted a variety of new multispectral cameras focused on the Earth. The resolution of Earth-focused cameras improved with the launch of the Skylab in May 1973. Rosman received ERTS data and played a major role in collecting Skylab imagery. ERTS and Skylab information from Bill R., interview, October 30, 2021.
55. Rosman station director M. Gary Dennis in "Rosman Tracking Station Will Make Astronomical History Following OAO," *Transylvania Times*, March 31, 1966, 1, 6. For more information on specific experiments, orbital parameters and technical capabilities of NASA spacecraft, see NASA, "NSA-NSSDCA Master Catalog," https://nssdc.gsfc.nasa.gov/nmc/spacecraft/query.
56. Every second, thermonuclear reactions in the center of the sun fuse 600 million tons of hydrogen into 596 million tons of helium. The missing 4 million tons is turned into pure energy. This energy, generated inside the sun, rises through successively cooler and less dense layers of gas, finally emerges and can be released as solar flares.
57. "Rosman One Year Old," *Transylvania Times*, 4.

58. Vandenberg AFB, on the California coast, is used to launch U.S. satellites into polar orbits.
59. Ezell, *Data Book*, 3:172.
60. "Rosman Tracking Station Aids in Ozone Research," *Transylvania Times*, March 31, 1975, 13; "*OGO-4* Reaches Near Polar Orbit to Study Solar Activity Effects," *Aviation Week and Space Technology*, July 7, 1967, 30.
61. For more on Fairbanks, see Edmund C. Buckley, director of tracking and data acquisition, to NASA associate administrator, "FY1962 CoF Estimates for Project 3379, Rosman Data Acquisition Facility," LS, February 14, 1962, 3. For more on Quito and Santiago, see "Astronomical History Following OAO," *Transylvania Times*, 6.
62. Arthur C. Clarke was an English science and science fiction writer, educator, inventor and futurist. He cowrote the screenplay for Stanley Kubrick's influential 1968 film, *2001: A Space Odyssey*. For quote, see Wales, *ATS Report*, xiii, xv; Joe Collins, chief ATS engineer at Rosman, interview with the author, Brevard, NC, December 30, 2021.
63. The second eighty-five-foot-wide dish was originally funded as part of Rosman II to provide backup for the first eighty-five-foot-wide dish in case of "data saturation" or "catastrophic failure." Plans later evolved to give the second eighty-five-foot-wide dish a separate mission to support communications experiments associated with the ATS program. See Edmund C. Buckley, director of tracking and data acquisition, to NASA associate administrator, "Communication Satellite Ground Support Requirements Relative to Rosman II Station Plans," LS, January 24, 1963.
64. Collins, interview, December 30, 2021.
65. Technically, the first satellite placed in geosynchronous orbit was NASA's "semi-synchronous" communications satellite-2 (SYNCOM-2), an experimental communications satellite launched in August 1964. The international telecommunications satellite (INTELSAT), the first commercial geosynchronous satellite communications satellite was launched in April 1965.
66. "Eighth Anniversary," *Transylvania Times*, 8.
67. Ibid.
68. Michael Collins stayed aboard the command module *Columbia* while Armstrong and Aldrin explored the moon's surface. Prior to 1972, there were separate manned space flight and scientific satellite ground stations, but when it came to important manned missions, it was "all hands on deck," and Rosman frequently provided backup communications with ATS satellites.
69. But this communication, like the mission itself, didn't go entirely as planned. Rosman engineers were ordered to pull the plug on the live nationwide television

feed when astronaut Jack Swigert, who had a reputation of being a playboy, began to reveal details about how his live-in girlfriend washed his underwear (Elizabeth Howell and Kimberly Hickok, "Jack Swigert: *Apollo 13* Command Module Pilot," Space, April 8, 2020, www.space.com; Matt McGregor, "Collins Tracks 50 Years of History," *Transylvania Times*, November 6, 2016, 1).

70. INTELSAT, originally formed as the International Telecommunications Satellite Organization in 1964, was an intergovernmental organization that owned and managed communications satellites.
71. *TIROS-1*, launched in low Earth orbit in April 1960, was the first U.S. weather satellite.
72. Space News Roundup, "First Applications Satellite Launched," December 9, 1966, 3 (printed).
73. NASA, "Applications Technology Satellite Program."
74. *ATS-1* was launched on December 6, 1966; the failed *ATS-2* on April 6, 1967; and *ATS-3* in November 1967.
75. Nachman and Bartlett, *Check-Out Report*, 1–4.
76. "Rosman Tracking Station Supports the *Apollo XI* in Indirect, Important Ways," *Transylvania Times*, July 24, 1969, 1.
77. Nachman and Bartlett, *Check-Out Report*, 1–4.
78. "Eighth Anniversary," *Transylvania Times*, 8.
79. Hurricane Camille (August 14–22, 1969) was, at the time, the second most destructive hurricane to reach the United States. When Camille made landfall in Bay St. Louis, Mississippi, it was a category five hurricane with sustained wind speeds over 175 miles per hour and a storm surge of twenty-four feet. Despite Weather Service warnings, 259 people died, and property damage was in excess of $1.4 billion (nearly $10 billion in 2022).
80. Legendary NASA engineer and manager Wernher Von Braun was instrumental in developing the ATS program and, according to Clarke, had a special affinity for *ATS-6*. Von Braun reportedly visited the Rosman ground station to tour ATS support equipment and facilities. Joe Collins, chief engineer for ATS at Rosman, remembered, "He asked a lot of questions, was down to earth.…He was one of the best administrators NASA had." Von Braun went on to meet Clarke in India, where the ATS experiment offered health and educational television to the poorest of the country. For Clarke's quote, see Wales, *ATS Report*, xv.
81. "Most Complex Satellite Ever Developed for Space," *Aviation Week and Space Technology*, August 6, 1973, 45.
82. Cal Carpenter, "Satellite Aids Appalachia Medical Teaching," *Transylvania Times*, October 24, 1974, C1.

83. Nachman and Bartlett, *Check-Out Report*, 2–4.
84. In 1975, *ATS-6* received data from the *geodynamics experimental ocean satellite-3* (*GEOS-3*), an experimental geodetic (Earth-measuring) satellite that was launched in a polar orbit from the military's Vandenberg Air Force Base in California. *GEOS-3* measured the Earth's shape and dynamic behavior, including volcanic and Earth crust movements. *ATS-6* also tracked the *Nimbus-6* polar orbit. NASA called *ATS-6* a "prototype" of a new satellite that would be known as the tracking and data relay satellite (TDRS) (Ezell, *Data Book*, 3:342).
85. "Effort for *ATS-6* Backup Fails," *Aviation Week and Space Technology*, November 4, 1974, 43. This backup *ATS-6* is on display at PARI, on loan from the Smithsonian Institution.
86. Of the twenty-six manned space flight (MSF) and space tracking and data acquisition network (STADAN) ground stations that began operating in the 1950s and 1960s, twelve were closed by 1975. The combined MSF and STADAN network was renamed the Spaceflight Tracking and Data Network (STADN) and included several aircraft and three tankers that General Dynamics, in 1964, converted to instrumentation ships for the Apollo program (Corliss, *Histories*, 407; Ezell, *Data Book*, 3:424).
87. Ezell, *Data Book*, 3:424.
88. Tsiao, *Read You*, 216. By 1978, two TDRS satellites in geostationary orbit, with one backup, were expected to track and relay data from fifty spacecraft per day.
89. Nimbus started as a joint military/civilian meteorological satellite program but incurred delays in its development. The military could not wait and developed its own a low-altitude weather satellite that was needed by fighter pilots who were planning bombing runs in Vietnam. The CIA also quickly needed a reliable weather satellite for its CORONA photograph satellite to avoid wasting valuable film resources by imaging cloud-covered targets. Ultimately, Nimbus became an experimental platform that tested new sensors for weather and Earth resource research.
90. ERTS was renamed Landsat in 1975.
91. The intelligence community identified over fifty uses for ERTS sensors, including monitoring the spread of pollution and melting sea ice, surveying oceanographic and marine resources, assessing earthquake damage and much more. Rosman downloaded imagery from Landsat but forwarded it to Goddard SFC for processing. Bill R., former NASA and DoD employee, e-mail exchange with author, September 30, 2021.
92. Rosman was to be one of five TDRS orbital support stations. Ezell, "Table 6-32 Tracking and Data Acquisition Stations, 1969-1978, SP-4012," in *Data Book*, 3:427.

93. "Eighth Anniversary," *Transylvania Times*, 8.
94. Under cost pressures, the NASA contract allowed for four short duration extensions of three months each. "Tracking Station to Recruit Local Persons in New Program," *Transylvania Times*, May 3, 1973, 1; Bendix Corporation advertisement, "Join the Space Team!" *Transylvania Times*, May 3, 1973, 21.
95. According to the NSA, Rosman Tracking Station followed some three dozen different classes of orbiting spacecraft (not including ATS) during the NASA years. See NSA, "Rosman Tracks," 1.
96. Western Union placed its first communications satellite into geostationary orbit in April 1974. After 1970, weather data from *ATS-1* and *ATS-3* was collected by NOAA, and *ATS-1* was used periodically as a communications satellite by state, federal and public organizations ("Eighth Anniversary," *Transylvania Times*, 8).
97. Station employees belonging to the International Brotherhood of Electrical Workers eventually won a 15 percent wage increase from Bendix. In a letter, NASA officials told North Carolina congressman Roy Taylor that the strike would have no impact on the status of the station. See Tsiao, *Read You*, 227.
98. Ken Atchison, "NASA to Change Tracking and Data Acquisition Operations," NASA Press Release, December 7, 1979, 2. Though NASA had plans in February 1979 to close Rosman in 1982, the timetable for closure was later accelerated to 1981 after the decommissioning of ATS and observatory-class satellites. The remaining 119 Bendix employees assigned to Rosman were offered jobs elsewhere within the company. Thirty accepted the Bendix offer and were transferred to other locations, and thirty-four accepted positions with the site's next employer, the DoD. Fifty-five either declined employment with Bendix or did not become DoD employees and were terminated. See Tsiao, *Read You*, 228; NASA, "Daily Activities Report," February 23, 1981, 2.
99. By August 1980, the DoD was in discussions with NSA, the General Services Administration (GSA) and the U.S. Forest Service. The closure of Rosman saved NASA $4 million per year in operating costs. See Terrence Finn, NASA director of legislative affairs to Representative Don Fuqua, chairman of the Committee on Science, Technology, House of Representatives, LS, August 20, 1980.
100. The NSA's statement that Rosman tracked "thirty-six (different) orbiting spacecraft" does not include the ATS program, Apollo-Soyuz or other backup responsibilities for the manned spaceflight program. See NSA, "Rosman Tracks," 1. "Fifty different satellites, up to forty per day during the 1970s" quote in Tsiao, *Read You*, 222.
101. Data derived from "Eighth Anniversary," *Transylvania Times*, 8.

Act II. Spies: The National Security Agency Assumes Control (1981–95)

102. *Facility Brochure*, 2.
103. NBC News, *The Eavesdropping Wars*, November 12, 1986 (television broadcast). The National Security Agency is under the U.S. Department of Defense.
104. Roller, *Tracking Station*, 9.
105. David, *Spies and Shuttles*, 81.
106. NORAD, a joint U.S.-Canada organization, provides aerospace warnings, air sovereignty and protection for North America. This includes monitoring man-made objects in space and the detection, validation and warning of attacks against North America from either aircraft, missiles or space vehicles.
107. David, *Spies and Shuttles*, 82.
108. The National Aeronautics and Space Act of 1958 requires NASA to "make available to agencies directly concerned with national defense discoveries that have military value or significance" (NASA, "History").
109. Both NASA/Goddard and NRO have overlapping interests, as each is involved in the design, development, production and operation of spacecraft. Common interests and understanding of unique technologies and their applications also seemed to spill over into personnel practices. For example, in 2019, Christopher Scolese, the director of Goddard SFC, retired from NASA to become the director of the National Reconnaissance Office (NRO).
110. Levine, *Managing NASA*, 231.
111. Ibid., 212.
112. The NRO, originally staffed by air force, CIA and navy employees, is responsible for designing, building, launching and operating U.S. reconnaissance satellites. The CIA had primary responsibility over photographic satellites. The existence of the NRO was declassified in 1992. "DoD (NRO) presently is dependent on…NASA/Goddard to provide radiation estimates…useful for Corona (photographic) satellites," quoted in Herbert Scoville Jr., DD/NRO to D/NRO, "Subject: Radiation Belt Monitoring," L, October 22, 1962.
113. Rosman collected data from at least one of these geodetic satellites, *GEOS-3*.
114. "Compulsion to Publicize," quoted in Paul E. Worthman, "NASA's Reconnaissance Activities," *Memorandum for the Record*, STM, 23 April 1965. This seemed to be a recurrent theme. NASA laser communications experiment had DoD interest—and likely sponsorship—but here, NRO also complained that, "too much of NASA's research was being publicized." Harold Wheeler, NRO, to Myron Kreuger, NASA, L, 12 November 1975.
115. David, *Spies and Shuttles*, 275.
116. Collins, interview, December 30, 2021.

117. Ibid.
118. The *ATS-6* communications antenna was the largest antenna at the time. U.S. Government Interagency Survey Applications Coordinating Committee, "Summary of Major SACC Coordinating Activities: 9 to 12, May 1969," National Security Archive, May 21, 1969, 3, https://nsarchive2.gwu.edu/NSAEBB/NSAEBB509/docs/nasa_29.pdf. Large antennas continue to hold special interest for the NRO satellite architects, because their narrow communications footprints are more difficult to intercept. See U.S. Government Interagency Panel, "36th Meeting of the Unmanned Spacecraft Panel of the Aeronautics and Astronautics Coordinating Board, Summary Minutes, No. 36," National Security Archive, April 14, 1964, 3, https://nsarchive2.gwu.edu/NSAEBB/NSAEBB509/docs/nasa_36.pdf.
119. Experiment no. 649 was identified as "classified." See Engler, Nash and Strange, *User Experiments*, 2–119.
120. "*ATS-6* Backup Fails," *Aviation Week and Space Technology*, 17.
121. Robert Seamans Jr., NASA, to Albert Wheelon, CIA DDS&T, L, September 15, 1965, https://www.cia.gov/readingroom/docs/CIA-RDP71B00508R000100120012-1.pdf.
122. CIA, "Untitled Memorandum, 25 August 1965," https://www.cia.gov/readingroom/docs/CIA-RDP85B00803R000200010019-2.pdf.
123. "The CIA cannot afford to turn its back on NASA expertise in reaching intelligence judgments for the president on such an important national topic," quoted in Albert Wheelon, "Meeting with Dr. Seamans and General 'Bozo' McKee," *Memorandum for the Record*, February 2, 1965, https://www.cia.gov/readingroom/docs/CIA-RDP71B00508R000100120026-6.pdf.
124. NASA's manned spaceflight program relied on the optics from NRO's GAMBIT (KH-4/8) photographic satellites. See David, *NASA's Acquisition and Use*.
125. "Ideally suited…Goddard presently supports satellite missions similar to that anticipated by the NRO," quoted in Alexander H. Flax, director of NRO, to assistant secretary of the air force for financial management, "NRO Command and Control Communications System," TL, n.d., 2, https://www.nro.gov/Portals/65/documents/foia/declass/Archive/NARP/1968%20NARPs/SC-2018-00032_C05108048.pdf.
126. "*ATS-6* was "used to control low Earth orbit (Nimbus) satellite equipment through command data transmission and track near-Earth satellites." The geophysical satellite, *GEOS-3*, was also used in the *ATS-6* relay satellite experiment. See Engler, Nash and Strange, "Experiment no. 617," in *User Experiments*, 2–110.

127. Weather satellites orbit ahead of photographic satellites to identify cloud covered targets, helping these satellites to image only clear targets, thus making the best use of its limited resources (National Reconnaissance Office, "Memo for the Record: Subject Aviation Week Article, 'Weather Coverage,'" November 13, 1973, https://www.nro.gov/Portals/65/documents/foia/declass/NROStaffRecords/97.PDF). The CIA earlier considered using Nimbus as an intelligence collection platform. See Chief of CIA Systems Analysis staff, "'Industry Capabilities and a Proposal for Work,' Memorandum of Conversation," April 19, 1965, 1, https://www.cia.gov/readingroom/docs/CIA-RDP80B01138A000100010085-0.pdf.
128. Nachman and Bartlett, *Check-Out Report*, iv.
129. "It's widely believed that the tracking and data relay satellite (TDRS) system was built for civilian purposes, but there is evidence that National Security Agencies were key users, including a draft MOU from early 1979 by the DoD sent to NASA regarding the DoD's use of TDRS." See David, *Spies and Shuttles*, 85.
130. The transition in turning over the NASA site to the DoD/NSA was seamless. Just as NASA was vacating the property in January 1981, the NSA was given a special-use permit to begin operating at Rosman. This was prior to the NSA's establishment of an MOU with the U.S. Forest Service, outlining its planned operations for the site. See NASA, "Daily Activities Report," 2. The DoD's National Security Agency (NSA) is responsible for collecting, decoding and reporting our adversaries' electronic signals, including political and military communications, among a variety of other responsibilities.
131. Tsiao, *Read You*, 228.
132. Jo Ann Jackson, the spouse of NASA's second Rosman director, James C. "Chuck" Jackson, interview with the author, Brevard, NC, December 16, 2021.
133. Even after four decades, the NSA, CIA and NRO remain silent about activities that took place at Rosman. What follows is largely supposition based on the international context and events at the time of Rosman's acquisition by the NSA and data acquired through the National Archives, the Freedom of Information Act and declassified documents collected from the CIA, NSA and NRO websites and other unclassified sources.
134. Depending on how each antenna is directed, they are capable of collecting signals from a variety of fixed or mobile transmitters within their "beamwidths." NASA's Rosman antenna could collect signals from ground-, air-, sea- and space-based platforms.
135. According to the NSA, Rosman's "mission start" date was April and its "official start" date was July 1, 1981. The NSA was so anxious to use Rosman's

capabilities that it received permission to begin operations coincident with NASA's departure in January 1981. See the "official date" in NSA, "Rosman Tracks," 1. For information about the NSA's early use, see NASA, "Daily Activities Report," 2.

136. NSA, *Organizational Manual*, F-28. Don C. was Rosman's third chief of station.
137. Pete Zamplas, "Research Station Remains a Mystery," *Transylvania Times*, June 28, 1990, 14A.
138. Ibid.
139. Rosman also likely served as a ground station for NSA satellites that were intercepting foreign signals in western and far eastern geosynchronous and other orbits.
140. Rowe, interview, April 6, 2022. The DoD took over 220 acres from NASA, of which 70 were developed.
141. Johnson, "Cuban Brigade," in *American Cryptology*, 3:258; Turner, *Annual Report to Congress*, 18.
142. CIA, "Soviet Global Military Reach," 29–30.
143. Cuba was assessed to have forty-five thousand military personnel in developing countries. See CIA, "Cuban-Soviet Connection," iv.
144. "Farewell Ceremony Held for Rosman Tracking Station," *Transylvania Times*, November 2, 1975, 10A.
145. At the time, the only other intelligence signals station outside of the Soviet Union was located in Mongolia. "Greatest Asset" quoted in Russian-American Trust and Cooperation Act of 2000, 106th Cong., 2nd sess., 2000, H.R. 4118, 5, https://www.congress.gov/congressional-report/106th-congress/house-report/668.
146. DIA, "Cuba," 3.
147. Clarence A. Robinson Jr., "USSR Cuba Force Clouds Debate on SALT," *Aviation Week and Space Technology*, September 10, 1979, 16–17.
148. It appears that cost was an important factor in the Soviet's selection of Lourdes, too. Though the Soviets provided the Cubans a lot of economic and military support, they used Lourdes rent free until 1992.
149. *Sunday Times*, untitled, January 26, 1997, quoted in Russian-American Trust and Cooperation Act of 2000, 106th Cong., 2nd sess., 2000, H.R. 4118, 4, https://www.congress.gov/congressional-report/106th-congress/house-report/668.
150. This judgment was also made by NBC News in "The Eavesdropping Wars," television episode, which aired on November 12, 1986.
151. To maintain communications security, the Soviets normally only transmit to ground stations on their territory. See James D. Burke, "The Missing Link," *CIA*

Studies in Intelligence (Winter 1978): 1–10, https://www.cia.gov/readingroom/docs/CIA-RDP80-00630A000100050001-4.pdf.

152. This Air Force balloon radar surveillance system, SEEK SKYHOOK, located eighteen miles east, northeast of Key West operated twenty-four hours per day. See Robinson, "Force Clouds Debate," 16–17.

153. *Molniya-1* and *Molniya-3* are two different series of satellites. Though the *Molniya-1* series was first launched in 1966, the CIA believed that by the 1970s, it would mostly be used for military command and control communications. *Molniya-3* became operational for use by the Soviet military in the mid-1980s and "undoubtedly uses some channels for communicating between Moscow and Havana." *Molniya-3* hosted the "hotline" between Moscow and Washington, D.C. See CIA, "Soviet Statsionar Satellite Communications System," 14.

154. When the DoD departed Rosman, real estate sales brochures for the abandoned site listed the frequency ranges of the four satellite dishes the NSA left behind. The antennas were tuned to receive the downlink of Molniya satellites. According to CIA, *Molniya-1* used a portion of the "L-Band," between 0.975 and 1.0 GHz (downlink). Rosman's eighty-five-foot-wide dish was tuned to collect at 0.95 to 1.0 GHz. *Molniya-3* communicated at 3.65 to 3.9 GHz (downlink). Both eighty-five-foot-wide dishes and two others (forty- and fifteen-foot-wide dishes) were tuned at 3.4 to 4.2 GHz. See CB Commercial Real Estate, "Available Satelite [*sic*] Communications and Data Center For Sale," 1995, 3; CIA, "Soviet Statsionar Satellite Communications System," 85.

155. These dishes were evident in a photograph taken by the DoD that was printed in a publicly distributed real estate brochure (see on page 97). Shchelkovo: National Photographic Interpretation Center, "Possible Moscow Terminal for USSR-Cuba Satellite Communications Link Identified," June 1982, 1, https://www.cia.gov/readingroom/docs/CIA-RDP82T00709R000200780001-0.pdf.

156. Geostationary satellites eliminate "the operational break," which "represents a distinct improvement mainly for communications…(with) Havana." Quoted in "East Europe Report, Scientific Affairs," Foreign Broadcast Information Service, September 24, 1980, 3, 5.

157. Another Soviet geostationary satellite, *Ekran* ("Screen'), was first launched in geosynchronous orbit in 1976 but was positioned too far east of the meridian to be seen by Rosman's satellites. It probably provided civilian communications, television and military communications to the central and eastern USSR (Wikipedia, "*Ekran*," https://en.wikipedia.org/wiki/Ekran).

158. For every one geosynchronous satellite, Molniya needed four. Also, Molinya required complex ground equipment, including tracking and acquisition

antennas (Philip K. Klass, "NSA Jumpseat Program Winds Down," *Aviation Week and Space Technology*, April 2, 1990, 46).

159. Since Rosman (at 82 degrees west longitude) had prime responsibility for commanding NASA's *ATS-1* in geostationary orbit 68 degrees to its west (150 degrees west longitude), Rosman could likely access Soviet Gorizont geostationary communications satellites off the coast of Africa, 68 degrees to its east (14 degrees west longitude), as long as topography did not interfere with collection. Raduga, located 25 degrees west longitude, would have been an easier collection target. Rosman's longitude, like Lourdes, near Havana, is 82 degrees west longitude. Rosman would have had access similar to Lourdes to both Soviet Gorizont and Raduga geostationary satellites.

160. CIA director William Casey to Charles Z. Wick, director, U.S. Information Agency, L, September 5, 1986, https://www.cia.gov/readingroom/docs/CIA-RDP89-01147R000100080003-2.pdf. In the letter, Casey thanked Wick for an unclassified news broadcast that the USIA had just received through its recent access to Gorizont signals. Casey probably deflated the USIA director when he noted that the CIA had been collecting these same signals for more than a year. Casey also stated that the CIA had been collecting unclassified Molniya news broadcasts for some time prior to 1986.

161. The U.S. Congress reported the Soviet military operated Raduga channels between 7.25 and 7.75 GHz. Rosman's eighty-five-foot-wide antenna was also tuned to receive signals between 7.2 and 7.8 GHz. For information on Raduga frequencies, see Congress, U.S. Senate, "Soviet Space Programs," 322. For information on Rosman frequencies, see CB Commercial Real Estate, "Available Satelite [*sic*]," 3.

162. The Gorizont had multiple antennas tuned to frequencies in the "C band" between 3.65 and 3.925 GHz and 11.525 GHz in the "KU band." One Rosman eighty-five-foot-wide and one forty-foot-wide antenna were tuned to collect "C band" signals at 3.4 to 4.2 GHz and 11.5 to 11.7 GHz in the "KU band." For information on Gorizont frequencies, see Frequency Plan Satellites, www.frequencyplansatellites.altervista.org. For information on Rosman frequencies, see CB Commercial Real Estate, "Available Satelite [*sic*]," 3.

163. The DoD has been silent about its early use of commercial communications satellites, but it was likely a key user of TDRS. See David, *Spies and Shuttles*, 85. During Desert Shield/Desert Storm (August 1990–February 1991), about 24 percent of military communications into and out of the operations' theaters went through commercial satellites (U.S. Space Command, "Operation Desert Shield and Desert Storm," 4).

164. Cited earlier: see the top image on page 57; Ezell, *Data Book*, 3:427: "As of 1989, about 10 percent of nonsecure U.S. military communications were transmitted via U.S. civilian telecommunications networks." See CIA, "Intersputnik's Competitiveness," 13. This included TDRS-1, which malfunctioned after its launch from the space shuttle *Challenger* in April 1983 and was first placed off the coast of Brazil (41 degrees west longitude), and TDRS-3 (1988) over the Venezuela-Brazil border (62 degrees west longitude). Given Rosman's past responsibilities, it likely collected data from the easily accessible but malfunctioning TDRS-1 satellite. Of the four satellite dishes that the DoD did not remove from the site, two receivers (both eighty-five-foot-wide dishes) could collect signals at 2.2–2.35 GHz, a portion of the known TDRS "S band" (2.0–2.3 GHz).
165. The malfunctioning TDRS-1 satellite was moved from 41 to 67 degrees west longitude, almost directly over Grenada, a few months before the United States invaded the island nation in October 1983.
166. Stella Trapp, "What Signals Are Being Received at Rosman?" *Transylvania Times*, November 11, 1985, 1.
167. U.S. Army Corps of Engineers, *Tracking Facility, Rosman*.
168. Don Cline, president, Pisgah Astronomical Research Institute, interview with the author, at PARI, April 7, 2022.
169. Trapp, "What Signals?" 1.
170. Cline, interview, April 7, 2022.
171. The Central Security Service (CSS), composed of elements of five military cryptologic components, ensures "timely, accurate cryptologic support, knowledge and assistance to the military cryptologic community." See NSA, "Central Security Service," https://www.nsa.gov/about/central-security-service/. Rosman was a field station that had signal research responsibilities and opportunities to respond to ad hoc tasking for immediate high-priority collection. The NSA stated there were three primary functions of the Rosman site. All three were redacted in the near thirty-year-old document. See NSA, *Organizational Manual*, F-26–27.
172. NSA, *Organizational Manual*, F-26.
173. What follows is largely supposition and is based on the international events at the time of Rosman's acquisition by the NSA and on data acquired through the National Archives, the Freedom of Information Act and declassified documents collected from the CIA, NSA, NRO websites and other unclassified sources. An NBC news broadcast, "The Eavesdropping War," aired on November 12, 1986, and offered a few similar, though largely unsubstantiated, conclusions.
174. "Four huge satellite dishes." See NSA, "Rosman Tracks," 1.

175. There were more than ten vacant concrete antenna pads. The DoD made public a photograph of the facility that showed most of the antennas (see the top image on page 73). See Cline, interview, April 7, 2022.
176. Of note, high frequency capabilities used with the *ATS-6* program (at 20 and 30 GHz and above) would have been available to NSA, but nearly all these antennas were removed after the agency abandoned the site.
177. Encryption is a technique used to mask communications. According to CIA documents, Lourdes reportedly used both open (unencrypted) and encrypted communications: "The Soviets didn't do all they could to conceal the brigade… using landlines or enciphered communications, only," quoted in NPIC, "Establishment of the Cuban Analytic Center, NPIC's Current Efforts in Cuba, Q&A," September 9, 1979, 1, https://www.cia.gov/readingroom/docs/CIA-RDP81B00401R000200060006-1.pdf.
178. NSA, *Organizational Manual*, F-27–28.
179. Chinese communications also could have been collected from U.S. military and intelligence geosynchronous satellites within Rosman's field of view in the eastern Pacific.
180. "Using satellites and high-speed computers, [the Lourdes base] can pick up millions of microwave transmissions every day and communicate with Russian spies operating on the American continent.…It would appear that the Lourdes facility plays an important role in supporting such espionage aimed against the United States." *Sunday Times*, January 26, 1997, quoted in Russian-American Trust and Cooperation Act of 2000, 106th Cong., 2nd sess., 2000, H.R. 4118, 4, https://www.congress.gov/congressional-report/106th-congress/house-report/668.
181. See Congress, U.S. Senate, "Soviet Space Programs," 323.
182. CIA, "Statsionar," 15.
183. Ibid.
184. The SS-25 had a common design to the SS-20 and likely possessed similar redundant forms of command and control communications.
185. White House, "Strategic Forces," 4.
186. "Highest priority," quoted in Johnson, *American Cryptology*, 4:338.
187. The SS-20 missile system used redundant communications systems that included a troposcatter antenna, which bounced communications waves off the lower level of the atmosphere, and a special satellite communications vehicle that likely communicated with geostationary satellites. The SS-25 probably had these redundant communications, too. See CIA, "SS-20 IRBM Equipment Update," 1.
188. In 1986, legislation was introduced in the Senate specifically to authorize the secretary of defense to procure and install cryptographic equipment at domestic

satellite facilities to prevent the Soviets from intercepting U.S. communications. A Bill to Authorize the Procurement and Installation of Cryptographic Equipment at Satellite Communications Facilities Within the United States, 99th Cong., 1st sess., S.2348, April 22, 1986, https://www.congress.gov/bill/99th-congress/senate-bill/2348?r=39&s=1.

189. Zamplas, "Station Remains a Mystery," 14A.

190. Unclassified certificates of appreciation issued by the NSA thanking Rosman Research Station employees for their work during Desert Shield/Desert Storm suggest the station provided support to the field through the TDRS and/or the military's Defense Satellite Communications Systems (DSCS) satellites located in the eastern Atlantic. Both operations also used multispectral imagery from Landsat, which Rosman collected during its NASA years. See U.S. Space Command, "Operation Desert Shield and Desert Storm," 17, 41.

191. For more information on NSA's Central Security Service (CSS), see https://www.nsa.gov/about/central-security-service/.

192. In response to a Freedom of Information Act request questioning the NRO about its presence at Rosman Research Station, the NRO stated in its denial letter that it could "neither confirm nor deny the existence or nonexistence of records responsive to [the author's] request." NRO FOIA public liaison Anita Casamento to the author, TLS, April 7, 2021.

193. While the four satellite dishes left behind by the DoD were receive-only, Rosman undoubtedly had the ability to use these antennas to transmit and issue commands to U.S. intelligence satellites.

194. NSA, *History Today*, April 14, 2005, 1.

195. Reportedly, just as the last bolt was being tightened, the station chief received a call that Rosman would be closing. The dish was then dismantled and moved to the United Kingdom at a cost of $5 million. The presence of freshly poured concrete pads suggest the NSA had plans for an additional six twelve-meter-wide antennas on a ridge near the helipad (Cline, interview, April 7, 2022).

196. NSA stations are usually tenants on military bases or are hosted by foreign governments. Rosman Research Station was the first closure of an "NSA-owned" facility. See NSA, "Rosman Closes," 1.

197. NSA, "Cryptologic Site Closure Briefing," 2.

198. "Before the site's last mission was shut down in November 1994, word of the closure was revealed by the local press, motivating various state and local groups to begin making requests," quoted in NSA, "Rosman Closes," 1.

199. The phrase the NSA used to describe the reason for the site's closure is unusual. It said Rosman's "last mission was shut down," suggesting an alternative explanation—that the site's antennas were used to collect signals

from another U.S. or foreign satellite that had ended its useful life. See NSA, "Rosman Closes," 1.

200. The treaty barred signatories from deploying more than 6,000 nuclear warheads and a total of 1,600 intercontinental ballistic missiles (ICBMs) and bombers.

Act III. Enter the Pisgah Astronomical Research Institute (1998–Present)

201. NSA, "Cryptologic Site Closure Briefing," 1.
202. NSA, "Rosman Closes," 1. This represents about 750,000 sheets of paper that, if stacked, would exceed the height of a twenty-one-story building.
203. *Final Report and Proposal*, 3.
204. NSA, "Rosman Closes," 1. "Texas Joint Ventures" reportedly included a former NSA director on its board. The NSA believed the arrangement was a "done deal" and called the "sale," worth between $10 and $15 million, "a windfall for the Forest Service." This proved wishful thinking. Soon thereafter, the arrangement fell through. See NSA, "Rosman Tracks," 2.
205. Littlejohn, *Updated Report*, 3.
206. What was known as science education later evolved into a concept known as "STEM," a broader term that includes the related disciplines of science, technology, engineering and mathematics.
207. The U.S. Forest Service was also leasing two ranger stations in Marion and Burnsville that Cline purchased and traded for what became the PARI site. Rowe, interview, April 6, 2022.
208. Cline later received an in-person visit from a very angry realtor who was hoping to close the deal for the developer. "You cost me millions of dollars in commissions," the realtor said. Don had no sympathy, telling the realtor the "highest and best" use of the property was for the public's education and enjoyment (Cline, interview, April 7, 2022).
209. Don purchased a six-month "special use" permit to preserve and take the site out of mothballs before final approval was received by Congress. Cline remembers one of his first tasks was to show up at the front gate, his pickup truck groaning under the weight of seven fifty-five gallon drums filled with transformer fluid, and replace the fluid in existing transformers that didn't meet Environmental Protection Agency standards (Cline, interview, April 7, 2022).
210. John Avant, phone interview with author, April 9, 2022.
211. Chief Technology Officer Lamar Owen, interview with author at PARI, Balsam Grove, NC, February–June 2020.

212. Castelaz said that because his title yielded the acronym DASE (pronounced "Daisy"), he was ribbed mercilessly by his peers. Brevard college professor Mike Castelaz, interview with the author, Brevard, NC, February 28, 2022.
213. Lynley Hargreaves, "Issues and Events: Spy Station Retooled into Astronomy Institute," *Physics Today*, 2001, 2, https://physicstoday.scitation.org/doi/10.1063/1.1366064.
214. Christiansen, "Cold War Relic."
215. Robert Hayward, interview with the author, Brevard, NC, February 7, 2022.
216. Matt Shelby, letter to the author, "PARI Book Project," TL (email), April 12, 2022.
217. Christi Whitworth, phone interview with the author, May 17, 2022.
218. See Elizabeth Landau, "What the Obsolete Art of Mapping the Skies with Glass Plates Can Still Teach Us," *Smithsonian Magazine*, April 2019, https://www.smithsonianmag.com/science-nature/obsolete-art-mapping-skies-glass-plates-can-still-teach-us-180971890/.
219. Don Cline first recognized the importance of what was being lost when, in 1996, he heard that the admiral in charge of the U.S. Naval Observatory in Washington, D.C., threw out a highly valuable set of three hundred glass plates that, one hundred years earlier, recorded the solar eclipse of August 1896. The event reportedly cost the U.S. government $100,000 to record. Apparently, the collection was destroyed to make room for a new employee. Don Cline, president, Pisgah Astronomical Research Institute, interview with the author, at PARI, May 12, 2022.
220. To participate in this effort called Project Scope, visit PARI, "Project Scope," https://www.pari.edu/research/scope/.
221. In recent years, Don visited Harvard's Center for Astrophysics, where he learned about and rescued glass plates that were found in fourteen leaf lawn bags stored in a local barn. Don Cline, president, Pisgah Astronomical Research Institute, interview with the author, at PARI, May 19, 2022.
222. Zac Engle, phone interview with the author, April 22, 2022.
223. Tim DeLisle, interview with author, PARI, Balsam Grove, NC, February 23, 2020.
224. Engle, interview, April 22, 2022.
225. An International Dark Skies Park "is land possessing an exceptional or distinguished quality of starry nights that is specifically protected for its scientific, natural, educational, cultural heritage, and public enjoyment." See Dark Sky, "International Dark Sky Parks," https://www.darksky.org/our-work/conservation/idsp/parks/.

SELECTED BIBLIOGRAPHY

Central Intelligence Agency. "'The Cuban-Soviet Connection: Costs, Benefits and Directions.' Intelligence Assessment." April 1986. https://www.cia.gov/readingroom/docs/CIA-RDP04T00794R000100080001-0.pdf.

———. "'Increasing Intersputnik's Competitiveness: Motives, Prospects, and Implications.' Intelligence Assessment." October 1989. https://www.cia.gov/readingroom/docs/CIA-RDP90R00762R000300120001-5.pdf.

———. "'Soviet Global Military Reach.' National Intelligence Estimate (NIE) 11-6-84." April 1985. https://www.cia.gov/readingroom/docs/CIA-RDP87T00126R000400450001-6.pdf.

———. "'The Soviet Statsionar Satellite Communications System: Implications for INTELSAT.' Interagency Intelligence Memorandum." April 1976. https://www.cia.gov/readingroom/docs/DOC_0000283805.pdf.

———. "'SS-20 IRBM Equipment Update.' Reference Aid." August 1984. https://www.cia.gov/readingroom/docs/CIA-RDP91T01115R000100250003-5.pdf.

Christiansen, Wayne. "Cold War Relic Finds New Use as National/International Observatory." Asheville: University of North Carolina Press Conference, October 20, 2003. https://www.youtube.com/watch?v=CQeNFvN3gj8.

Corliss, William R. *The Evolution of the Satellite Tracking and Data Acquisition Network (STADAN)*. Greenbelt, MD: NASA, 1967. https://ntrs.nasa.gov/citations/19670008308.

———. *Histories of the Space Tracking and Data Acquisition Network (STADAN), the Manned Space Flight Network (MSFN) and the NASA Communications Network*

(NASCOM). Washington, D.C.: NASA, 1974. https://ntrs.nasa.gov/citations/19750002909.

David, James E. *NASA's Acquisition and Use of Classified Technologies in Its Lunar Exploration Program*. Washington, D.C.: National Security Archive, George Washington University, 2015. https://nsarchive2.gwu.edu/NSAEBB/NSAEBB509/.

———. *Spies and Shuttles: NASA's Secret Relationships with CIA and DoD*. Gainesville, FL: National Air and Space Museum with the University of Florida Press, 2015.

Defense Intelligence Agency. "'Cuba: Soviet Military Activities.' Intelligence Appraisal." September 26, 1978. https://www.cia.gov/readingroom/docs/CIA-RDP06T01849R000100030037-2.pdf.

Engler, Nicholas A., John F. Nash and Jerry D. Strange. *Applications Technology Satellites and Communications Technology Satellite User Experiments: 1967–1980*. Cleveland, OH: NASA-Lewis Research Center, August 1980. https://ntrs.nasa.gov/citations/19810003628.

Ezell, Linda Neuman. *Historical Data Book*. Vol. 3, *Programs and Projects 1969–1978*. Washington, D.C.: NASA, 1988. https://history.nasa.gov/SP-4012/vol3/sp4012v3.htm.

Final Report and Proposal for Assessment of the Rosman Research Station for Economic Development in Western North Carolina, for Consideration of DoD, the U.S. Forest Service and Army Corps of Engineers. Raleigh: North Carolina State University, Western North Carolina Regional Economic Development Commission, North Carolina Alliance for Competitive Technologies and Microelectronics Center of North Carolina, March 15, 1995.

Johnson, Thomas R. *American Cryptology During the Cold War, 1945–1989*. Book 1, *The Struggle for Centralization*. Fort Meade, MD: Center for Cryptologic History, NSA, 1995. https://nsarchive2.gwu.edu/NSAEBB/NSAEBB260/nsa-1.pdf.

———. *American Cryptology During the Cold War, 1945–1989*. Book 2, *Centralization Wins*. Fort Meade, MD: Center for Cryptologic History, NSA, 1995. https://www.nsa.gov/portals/75/documents/news-features/declassified-documents/cryptologic-histories/cold_war_ii.pdf.

———. *American Cryptology During the Cold War, 1945–1989*. Book 3, *Retrenchment and Reform*. Fort Meade, MD: Center for Cryptologic History, NSA, 1998. https://www.nsa.gov/portals/75/documents/news-features/declassified-documents/cryptologic-histories/cold_war_iii.pdf.

———. *American Cryptology During the Cold War, 1945–1989*. Book 4, *Cryptologic Rebirth*. Fort Meade, MD: Center for Cryptologic History, NSA, 1999. https://nsarchive2.gwu.edu/NSAEBB/NSAEBB426/docs/2.American%20

Cryptology%20During%20the%20Cold%20War%201945-1989%20Book%20 IV%20Cryptologic%20Rebirth%201981-1989-1999.pdf.

Levine, Arnold S. *Managing NASA in the Apollo Era.* Washington, D.C.: NASA, 1982. https://history.nasa.gov/SP-4102.pdf.

Littlejohn, M.A. *Updated Report on the Rosman Research Station.* Raleigh: North Carolina State University, May 29, 1997.

Nachman, M., and R. Bartlett. *Applications Technology Satellite* ATS-6 *in Orbit Check-Out Report.* Greenbelt, MD: NASA, Goddard Spaceflight Center, August 1974. https://ntrs.nasa.gov/api/citations/19740024205/downloads/19740024205.pdf.

National Aeronautics and Space Administration. "Applications Technology Satellite Program." May 22, 2016. https://science.nasa.gov/missions/ats.

———. "History Division Archives." https://history.nasa.gov/books.html.

———. "Key NASA Tracking Site to be Dedicated at Rosman, N.C." Press release, no. 63-240, October 24, 1963.

———. "Space Science Data Coordinated Archive." https://nssdc.gsfc.nasa.gov/nmc/spacecraft/display.action?id=1972-065A.

National Security Agency. "Cryptologic Site Closure Briefing to the NSA Board of Directors." Action memorandum, June 29, 1993.

———. *NSA/CSS Organizational Manual.* Fort Meade, MD: NSA, September 1, 1993.

———. "Rosman Closes Its Doors." *Communicator* 3, no. 17 (April 24, 1995): 3–4. https://www.governmentattic.org/docs/NSA_Tracking_Rosman-NC-NSA-Site.pdf.

———. "Rosman Tracks on to the End." In *Cryptologic Almanac.* Fort Meade, MD: Center for Cryptologic History, August 14, 1996. https://media.defense.gov/2021/Jun/29/2002751815/-1/-1/0/ROSMAN_TRACKS_TO_THE_END.PDF.

Tsiao, Sunny. *Read You Loud and Clear.* Washington, D.C.: NASA, 2008. https://history.nasa.gov/STDN_082508_508%2010-20-2008_part%201.pdf.

Turner, Stansfield. *Director of Central Intelligence Annual Report to Congress.* Washington, D.C.: CIA, January 29, 1980. https://www.cia.gov/readingroom/docs/CIA-RDP83M00171R002100110001-1.pdf.

U.S. Army Corps of Engineers. *Tracking Facility, Rosman, North Carolina.* Cary, NC: Savanah Engineering District Real Estate Office, April 8, 1981.

U.S. House of Representatives. "Tracking Site, 1 Announcement: House Committee on Science and Aeronautics, Aeronautical and Astronautical Events of 1961, 87th Cong., 2nd sess., Washington, D.C., June 7, 1962." https://history.nasa.gov/AAchronologies/1961.pdf.

U.S. Senate. "Soviet Space Programs: 1981–87. Committee on Commerce, Science and Transportation, 101st Cong., 1st sess., S. Washington, D.C., (Part 2), April 1989." https://www.google.com/books/edition/Soviet_Space_Programs_Space_science_spac/_ra3AAAAIAAJ?hl=en&gbpv=1&dq=Soviet+Space+Programs:+1981-87,"+Committee+on+Commerce,+Science+and+Transportation,+101st+Cong.,+1st+sess.,+S.+print+101-32+(Part+2)+April+1989&pg=PA267&printsec=frontcover.

U.S. Space Command. "Operation Desert Shield and Desert Storm." Assessment, January 1992.

Wales, Robert O., ed. ATS-6 *Final Engineering Performance Report.* Vol. 4. Greenbelt, MD: Goddard Space Flight Center, 1981. https://ntrs.nasa.gov/api/citations/19820008277/downloads/19820008277.pdf.

Wallace, Lane E. *Dreams, Hopes and Realities. NASA's Goddard Space Flight Center: The First Forty Years.* Washington, D.C.: NASA, 1999. https://history.nasa.gov/SP-4312/sp4312.htm.

White House. "Strategic Forces Modernization." *National Security Decision Directive* no. 178, (July 10, 1985): 4. https://www.cia.gov/readingroom/docs/CIA-RDP88B00443R000903780010-7.pdf.

INDEX

T

U

V

W

Y

ABOUT THE AUTHOR

Photograph courtesy of Luanne Allgood.

Craig Gralley, a former CIA senior executive, served as an analyst, manager and chief speechwriter for three agency directors. He's now a freelance writer and the author of *Hall of Mirrors–Virginia Hall: America's Greatest Spy of WWII*, a *Kirkus Reviews* "Best Book." His work has been published in the *Washington Post*, *WWII Magazine*, *Élan* and the *Sun*, among others. Craig graduated with honors from Allegheny College in Meadville, Pennsylvania, and holds master's degrees from Georgetown University (government) and Johns Hopkins University (writing). When he's not writing, Craig enjoys running and adventure traveling with his wife, Janet. Their son, Will, is a business owner and professional DJ in Washington, D.C. For more information about Craig and his work, visit www.craiggralley.com.